Laser-Based Nano Fabrication and Nano Lithography

Laser-Based Nano Fabrication and Nano Lithography

Special Issue Editors

Koji Sugioka
Ya Cheng

MDPI • Basel • Beijing • Wuhan • Barcelona • Belgrade

MDPI

Special Issue Editors
Koji Sugioka
RIKEN Center for Advanced Photonics
Japan

Ya Cheng
East China Normal University
China

Editorial Office
MDPI
St. Alban-Anlage 66
4052 Basel, Switzerland

This is a reprint of articles from the Special Issue published online in the open access journal *Nanomaterials* (ISSN 2079-4991) in 2018 (available at: https://www.mdpi.com/journal/nanomaterials/special_issues/Laser_Nano_Fabrication_Lithography)

For citation purposes, cite each article independently as indicated on the article page online and as indicated below:

LastName, A.A.; LastName, B.B.; LastName, C.C. Article Title. *Journal Name* **Year**, *Article Number*, Page Range.

ISBN 978-3-03897-410-9 (Pbk)
ISBN 978-3-03897-411-6 (PDF)

Cover image courtesy of Yongfeng Lu.

Contents

About the Special Issue Editors

Koji Sugioka, Dr. of Engin, received his B. S., Ms. Eng., and Dr. Eng. degrees in electronics from Waseda University (Japan) in 1984, 1986, and 1993, respectively. He Joined RIKEN in 1986 and is currently a Team Leader of Advanced Laser Processing Research Team at RIKEN Center for Advanced Photonics. He is concurrently a guest professor at Osaka University Tokyo Denki University. He was awarded the degree of Doctor Honoris Causa from University of Szeged, Hungary in 2018. He has made important contribution to both fundamental research on laser–matter interactions and diverse applications including practical applications in the said area. He is internationally renowned for his works on laser micro and nano processing, particularly ultrafast laser processing technology. He is currently a member of the board of directors of the Laser Institute of America (LIA) and Japanese Laser Processing Society (JLPS), a council member of the Intl. Academy of Photonics and Laser Engineering (IAPLE), and a Fellow of SPIE, OSA, LIA, and IAPLE. He is also an editor-in-chief of *Journal of the Laser Micro/Nanoengineering (JLMN) and an editor of Opto-Electronic Advances (OEA) and Advanced Optical Technologies (AOT)*.

Ya Cheng, PhD, received his B. S. from Fudan University in 1993 and his PhD in optics from Shanghai Institute of Optics and Fine Mechanics, Chinese Academy of Sciences in 1998. He is currently a professor at the Shanghai Institute of Optics and Fine Mechanics and Dean of the School of Physics and Materials Science, East China Normal University. His research interests include ultrafast nonlinear optics, femtosecond laser micromachining, and lithium niobate photonics. He is currently a council member of the Chinese Optical Society and a Fellow of IOP (UK). He is also an editor of *Journal of the Laser Micro/Nanoengineering (JLMN) and an editor of Micromachines*.

Preface to "Laser-Based Nano Fabrication and Nano Lithography"

The improvement of fabrication resolutions is an eternal challenge for miniaturizing and enhancing the integration degrees of devices. Laser processing is one of the most widely used techniques in manufacturing due to its high flexibility, high speed, and environmental friendliness. The fabrication resolution of laser processing is, however, limited by its diffraction limit. Recently, much effort has been made to overcome the diffraction limit in nano fabrication. Specifically, combinations of multiphoton absorption by ultrafast lasers and the threshold effect associated with a Gaussian beam profile provide fabrication resolutions far beyond the diffraction limit. The use of the optical near-field achieves nano ablation with feature sizes below 100 nm. Multiple pulse irradiation from the linearly polarized ultrafast laser produces periodic nanostructures with a spatial period much smaller than the wavelength. Unlimited diffraction resolutions can also be achieved with shaped laser beams. In the meanwhile, lasers are also widely used for the synthesis of nano materials including fullerenes and nano particles.

In view of the rapid advancement of this field in recent years, this Special Issue aims at introducing the state-of-the-art in nano fabrication and nano lithography, based on laser technologies, by leading groups in the field. Specifically, this Special Issue consists of two invited feature articles and ten contributed articles covering relevant topics of the laser-based nano fabrication and nano lithography, written by internationally recognized experts in the field. It includes the formation of periodic surface nano structures via the irradiation of intense ultrashort pulses, surface nano structuring below the diffraction limit using the optical near-field, three-dimensional micro-and nano-scale additive manufacturing based on two-photon polymerization, three-dimensional micro and nano fabrication inside transparent materials based on multiphoton absorption using ultrafast lasers, higher energy density confinement by Bessel ultrashort pulses for micro and nano fabrication, precision machining by liquid-assisted femtosecond laser ablation, the synthesis of functional nano particles by laser ablation in liquids, the formation of shape- and size-controlled metallic nanoparticles by pulsed-laser deposition and laser-induced dot transfer, and dual THz wave and X-ray generation by femtosecond laser excitation.

I believe that this Special Issue offers a realistic and comprehensive review of the state-of-the-art in nano fabrication and nano lithography and is beneficial for many researchers including students and young scientists working not only in the field but also in many other fields.

Last but not least, I would like to thank all of the article authors for their great effort and wonderful work to complete this informative Special Issue.

Koji Sugioka, Ya Cheng
Special Issue Editors

nanomaterials

MDPI

Article

Nanostructure Formation on Diamond-Like Carbon Films Induced with Few-Cycle Laser Pulses at Low Fluence from a Ti:Sapphire Laser Oscillator

Seiya Nikaido, Takumi Natori, Ryo Saito and Godai Miyaji *

Department of Applied Physics, Tokyo University of Agriculture and Technology, 2-24-16 Nakacho, Koganei, Tokyo 184-8588, Japan; nikaidada573@gmail.com (S.N.); impossible.is.nothing.1111@gmail.com (T.N.); s188097x@st.go.tuat.ac.jp (R.S.)
* Correspondence: gmiyaji@cc.tuat.ac.jp; Tel.: +81-42-388-7153

Received: 2 June 2018; Accepted: 14 July 2018; Published: 16 July 2018

Abstract: This study reports the results of experiments on periodic nanostructure formation on diamond-like carbon (DLC) films induced with 800 nm, 7-femtosecond (fs) laser pulses at low fluence from a Ti:sapphire laser oscillator. It was demonstrated that 7-fs laser pulses with a high power density of 0.8–2 TW/cm^2 at a low fluence of 5–12 mJ/cm^2 can form a periodic nanostructure with a period of 60–80 nm on DLC films. The period decreases with increasing fluence of the laser pulses. The experimental results and calculations for a model target show that 7-fs pulses can produce a thinner metal-like layer on the DLC film through a nonlinear optical absorption process compared with that produced with 100-fs pulses, creating a finer nanostructure via plasmonic near-field ablation.

Keywords: femtosecond laser; laser ablation; nanostructure formation; surface plasmon polaritons; near-field; diamond-like carbon

---

1. Introduction

Superimposed femtosecond (fs) laser pulses can form a periodic nanostructure (PNS) on solid surfaces through ablation, where the period size d is typically 10–20% of the laser wavelength λ [1–6]. There has been considerable interest in this surface phenomenon for application in laser nanoprocessing, beyond the diffraction limit of light. Numerous studies have been conducted to understand the mechanism responsible for PNS formation [7–10]. The experimental conditions and laser parameters for PNS formation have been identified for various target materials, and the dominant physical mechanisms responsible for nanostructuring have been determined.

Based on a series of experiments and model calculations, Miyazaki and Miyaji found that PNS formation is induced by fs laser pulses at a moderate fluence F through: a bonding structure change in the material [11–13]; generation of high-density electrons on the target surface, leading to the formation of a metal-like layer through linear and nonlinear optical absorption [13–15]; near-field ablation around the corrugated nanosurface [13–15]; and excitation of standing surface plasmon polariton (SPP) waves [9,10,15–17]. These laser–matter interaction processes can explain the origin and growth of PNSs on diamond-like carbon (DLC) [15], Si [16], GaN [10,17], Ti, and stainless steel [9], and theoretical calculations agree well with the observed nanoperiod, which is much smaller than $\lambda/2$. Based on the physical mechanism, control methods for the PNS shape have been developed, allowing the formation of homogeneous nanogratings [9,10,17] and a saw-like PNS [18] in air. However, some important processes for PNS formation are still unknown, and there is no consensus regarding the detailed mechanism.

For various kinds of material, it has been reported that the d value for a PNS increases with increasing F for the fs laser pulses at a fluence F of a few 100 mJ/cm^2 to a few J/cm^2 with a power density I of a few TW/cm^2 [7,8]. Previous studies have concluded that this increase is attributed to the increasing thickness of the metal-like layer produced on the target material with increasing F [16,17]. However, this has never been experimentally confirmed.

The proposed mechanism of PNS formation suggests that a thin metal-like layer can be produced by the fs pulses at low F with $I \sim$TW/cm^2 via a nonlinear absorption process, allowing confirmation of the thickness effect for nanostructuring. In this paper, we report the experimental results of PNS formation on DLC films irradiated with 800 nm, 7-fs laser pulses with a high power density I of 0.8–2 TW/cm^2 and a low fluence F of 5–12 mJ/cm^2 delivered from a laser oscillator. The results indicate the formation of a PNS with a period of $d = 60$–80 nm that *decreases* with increasing F. Based on the experimental results and a model calculation, it is shown that the excitation of SPPs at the interface between the thin metal-like layer and the DLC is certainly responsible for the nanostructuring process, and that the decrease of d is attributed to the decreasing wavelength of the SPPs with increasing F through an increase of electron density in the thin metal-like layer.

2. Experimental

Figure 1 shows a schematic diagram of the optical configuration used in the ablation experiments. As fs laser pulses with a high power density I at low fluence F can produce a thin metal-like layer on a target surface, the output of a Ti:sapphire laser oscillator was used in the experiments. The pulse duration $\Delta\tau$ was ~7 fs, the wavelength λ was 680–940 nm, the repetition rate f_{rep} was 80 MHz, and the pulse energy U_{pulse} was ~5 nJ. The pulses were so-called few-cycle laser pulses, where the electromagnetic field oscillates for a few cycles [19]. The temporal and spectral profiles of the fs pulses were monitored with a spectral phase interferometry for direct electric-field reconstruction (SPIDER) device and a spectrometer, respectively. When measuring the temporal profile, a silver mirror was inserted to propagate the pulses to the SPIDER device. The output just after the oscillator had a negative group delay dispersion, which was compensated for to minimize the pulse duration by passing the beam through a beam splitter (thickness: 1 mm) and a glass plate (thickness: 1 mm). The laser pulses were spatially expanded with a pair of convex and concave silver mirrors and focused onto the target surface with a ×40 Schwarzschild-type reflective objective (numerical aperture: 0.50) to a spot size w_0 of ~2 μm ($1/e^2$ radius) on the surface, since the group delay dispersion had to be suppressed to obtain laser pulses with a high power density. A CMOS camera was used to image the focused beam on the target surface. The pulse energy U_{pulse} just after the objective was measured with a pyroelectric detector, and the peak fluence $F = 2\,U_{pulse}/(\pi\,w_0^2)$ and the peak power density $I = F/\Delta\tau$ of the fs laser pulses on the target surface were estimated.

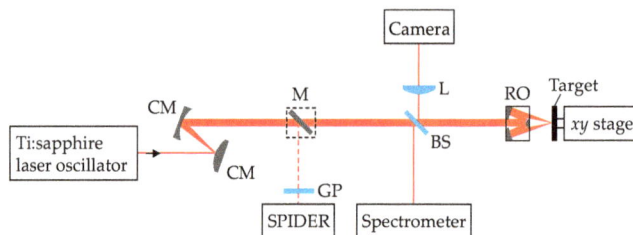

Figure 1. Schematic diagram of optical configuration for nanostructure formation. CM, convex or concave mirror; M, mirror; GP, glass plate; BS, beam splitter; L, lens; RO, reflective objective.

As the target, we used a DLC film (thickness: 1.7 μm) that was deposited on a polished silicon substrate with a plasma-based ion implantation system. The root-mean-square value of surface roughness was measured to be less than 1 nm with a scanning probe microscope (SPM). The target was set on an *xy* motorized stage, which could move at a constant speed v of 0.1–100 μm/s. The surface morphology was observed using a scanning electron microscope (SEM) and the SPM.

A two-dimensional Fourier transform was applied to the SPM images to analyze the distribution of the spatial periodicity in the surface structure along the polarization direction. The bonding structure of the target surface irradiated with the fs pulses was analyzed using micro-Raman spectroscopy with a diode-pumped, single-longitudinal-mode, 532 nm laser beam focused with a ×40 objective.

3. Results and Discussion

Figure 2a–c show SEM and SPM images and spatial frequency spectra of DLC films irradiated with 7 fs pulses with $I = 1$ TW/cm^2 at $F = 6$ mJ/cm^2 for $v = 0.1$–10 μm/s. For $v = 100$ μm/s, the surface was observed to swell and was not ablated because of the small shot number of the laser pulses onto the target surface. When v was decreased to 10 μm/s (i.e., the shot number increased), the formation of a PNS with a period d of ~50 nm was observed on the ablated DLC surface, as shown in Figure 2a. The line-like structure was perpendicular to the direction of polarization. When v was decreased to 1 μm/s, a PNS with d of ~70 nm formed, as shown in Figure 2b. With a further decrease of v to 0.1 μm/s, deeper ablation traces with d of ~80 nm formed, as shown in Figure 2c. For comparison, the target surfaces were also irradiated by 100-fs laser pulses with $I = 0.1$ TW/cm^2 at the same F. These pulses were produced by a glass plate (thickness: 3 mm) positioned just after the laser oscillator. As shown in Figure 2d, a PNS did not form on the ablated surface under these conditions.

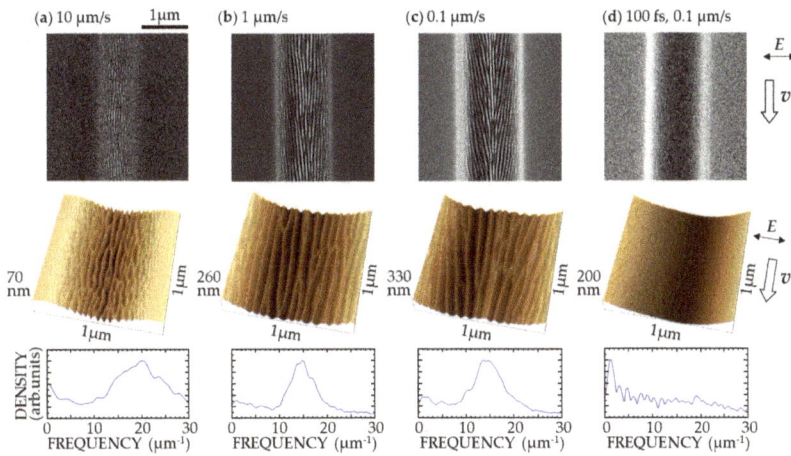

Figure 2. Scanning electron microscopy (SEM) images (**top**), scanning probe microscopy (SPM) images (**middle**), and spatial frequency spectra (**bottom**) of a diamond-like carbon (DLC) film surface irradiated with 7-fs pulses, with $I = 1$ TW/cm^2 at $F = 6$ mJ/cm^2 for (**a**) $v = 10$ μm/s, (**b**) $v = 1$ μm/s, and (**c**) $v = 0.1$ μm/s, and (**d**) those irradiated with 100-fs pulses with $I = 0.1$ TW/cm^2 at $F = 6$ mJ/cm^2 for $v = 0.1$ μm/s. *E* and *v* denote directions of polarization and laser scanning, respectively.

In previously reported experiments, PNSs formed on DLC films with 100-fs laser pulses with $I = 1$–2 TW/cm^2 at $F = 100$–200 mJ/cm^2, delivered from a chirp-pulse amplification Ti:sapphire laser system [3,11–15]. The results shown in Figure 2 suggest possible laser–matter interaction processes for PNS formation, as discussed in previous studies [13–16]. As v is decreased, a bonding structure change—from DLC to glassy carbon (GC)—is induced in the surface layer. This produces nanometer surface roughness due to swelling of the material, as a thin layer with a high electron density is produced on the surface through a nonlinear optical absorption process. On the highly curved swollen metal-like surface, an intense near-field is generated that enhances the incident electric field and initiates nanoscale ablation. Then, SPPs are transiently excited via coherent coupling of the incident laser pulses with the corrugated surface, where the GC layer, including high-density electrons, works as a thin metal layer between air and the DLC for the excitation of SPPs [20]. The periodic enhancement of the near-field of SPPs excited in the surface layer induces ablation, which forms a PNS on the surface. The experimental results shown in Figure 2 indicate that such a process occurs sufficiently when a DLC film is irradiated with 7-fs pulses with a high density of 1 TW/cm^2 at a low fluence of 6 mJ/cm^2.

An increase in F is expected to increase the density of the free electrons produced in the surface layer, leading to a change in surface morphology. To confirm this, surfaces were ablated with 7-fs pulses for $v = 0.1$ μm/s for $F = 5$–12 mJ/cm^2, corresponding to $I = 0.8$–2 TW/cm^2. The results are shown in Figure 3. At the lowest F, multiple shots produced a PNS with $d \sim 85$ nm; at the highest F, multiple shots produced a finer PNS with $d \sim 60$ nm. Figure 4 plots the d value obtained from the isolated peak position in the Fourier spectrum of the SPM images as a function of F and I. With increasing F, d decreases from about 85 to 60 nm. For irradiation with 100-fs laser pulses with $I = 1$–4 TW/cm^2 at $F = 100$–400 mJ/cm^2, it has been reported that the d value of the PNSs formed on various kinds of material (e.g., DLC, TiN, stainless steel, Ti, Si, and GaN) increased with increasing F [3,9,16,17], which is opposite to the results obtained in the present study. This suggests that low-fluence fs pulses with a high power density play a crucial role in the surface morphological change that leads to nanostructuring.

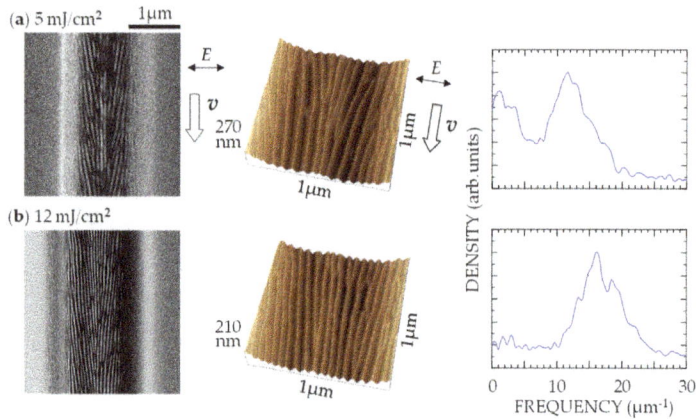

Figure 3. SEM images (**left**), SPM images (**center**), and spatial frequency spectra (**right**) of DLC film surface irradiated with 7-fs pulses at $v = 0.1$ μm/s for (**a**) $I = 0.8$ TW/cm^2, $F = 5$ mJ/cm^2 and (**b**) $I = 2$ TW/cm^2, $F = 12$ mJ/cm^2. E and v denote directions of polarization and laser scanning, respectively.

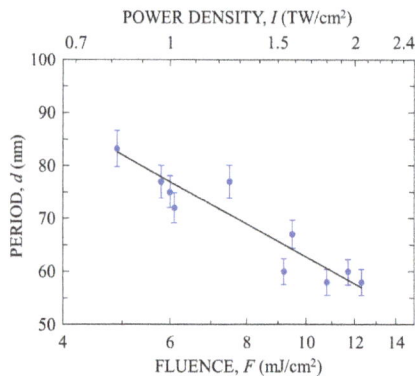

Figure 4. Period d of a periodic nanostructure (PNS) on DLC film formed with 7-fs laser pulses as a function of F and I for $v = 0.1$ μm/s.

In a previous study, we reported that PNS formation on a DLC surface is preceded by a change in the bonding structure, from DLC to GC [13]. The swelling of the target surface observed for $v = 100$ μm/s indicates that the onset of ablation at $v \leq 10$ μm/s is preceded by a change in the bonding structure to GC in the target surface. To confirm this, Raman spectra were obtained from surfaces ablated with 7-fs pulses with $I = 1$ TW/cm^2 at $F = 6$ mJ/cm^2 for $v = 0.1$ μm/s. The results are shown in Figure 5, together with spectra of surfaces ablated with 100-fs pulses with $I = 0.1$ TW/cm^2 at $F = 6$ mJ/cm^2 for $v = 0.1$ μm/s and non-irradiated DLC for comparison. Each spectrum is normalized to give a maximum intensity of unity. The asymmetric broad spectrum for the non-irradiated DLC has a single peak at 1530 cm^{-1}, which mainly consists of two spectra at peaks at ~1360 cm^{-1} (D band) and ~1590 cm^{-1} (G band) [21]. The D and G bands are attributed to bond angle disorder in sp^2 graphite-like micro/nanodomains and bond stretching between pairs of sp^2 atoms in both the rings and chains, respectively. The ratio of the intensities of the D and G peaks (I_D/I_G) and the position of the G peak have been reported to indirectly indicate the composition ratio of sp^2 and sp^3 bonding structures in DLC films [22–24]. These reports have shown that an increase in I_D/I_G and a shift of the G peak to a higher frequency represent an increase in the amount of sp^2 structures. The spectra from surfaces ablated with 7-fs and 100-fs pulses, shown in Figure 5, clearly show two spectral peaks at 1355 and 1590 cm^{-1}, respectively, indicating an increase in disordered carbon or GC [25–28]. As shown in Figure 5b, I_D for the surface irradiated with 7-fs pulses is smaller than that for the surface irradiated with 100-fs pulses. In addition, the position of the G peak for 7-fs pulses is shifted less than that for 100-fs pulses. These results show that less GC existed in the target surface irradiated with 7-fs laser pulses compared to that which existed with 100-fs pulses, despite the same F.

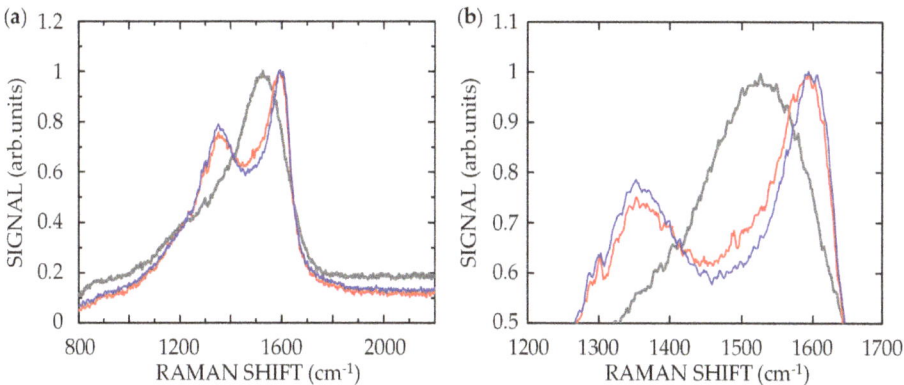

Figure 5. (a) Raman spectra of non-irradiated DLC film (gray) and DLC films irradiated with 7-fs (red) and 100-fs (blue) pulses at $F = 6$ mJ/cm^2 for $v = 0.1$ μm/s; (b) expanded spectra of (a) in the vicinity of the peaks of D and G bands.

To examine the the bonding structural change and ablation processes in detail, Raman spectra were obtained from a DLC film irradiated with 7 fs pulses with $I = 1$ TW/cm^2 at $F = 6$ mJ/cm^2 for various values of v ($v = 0.1–100$ μm/s). For comparison, spectra were also obtained from a film irradiated with 100 fs pulses with $I = 0.1$ TW/cm^2 at $F = 6$ mJ/cm^2. The peak intensities and positions of the D and G bands in the spectra were identified using a curve-fitting program with the Lorentzian function [29]. Figure 6a shows I_D/I_G plotted as a function of v. In the spectrum of the non-irradiated DLC film, I_D/I_G was ~1.25. For $v = 100$ μm/s, the ratio for both 7-fs and 100-fs pulses increased to ~1.5. With a decrease in v, the ratio monotonically increased, with that for 7-fs pulses being smaller than that for 100-fs pulses. Figure 6b shows the position of the G peak plotted as a function of v. In the spectrum of the non-irradiated DLC film, the G peak position was ~1582 cm^{-1}. For $v = 100$ μm/s, the position for both 7-fs and 100-fs pulses shifted to ~1590 cm^{-1}. With decreasing v, the position monotonically shifted to higher frequencies, with that for 7-fs pulses being at lower frequencies than that for 100-fs pulses. These results show two crucial processes for surface modification and subsequent ablation. For $v = 100$ μm/s, where both 7-fs and 100-fs pulses with the same F induced only swelling and no ablation on the target, the change in the spectra shown in Figure 6 indicates that the amount of GC at the surfaces irradiated with 7-fs and 100-fs pulses is the same, and that the surface phenomena do not depend on I. For $v \leq 10$ μm/s, where both 7-fs and 100-fs pulses with the same F induced not only a bonding structure change but also ablation on the target, the experimental results indicate that 7-fs pulses with higher I were strongly absorbed near the target surface through a nonlinear optical absorption process, forming a thinner GC layer than that produced by 100-fs pulses. The surface of the layer was then ablated.

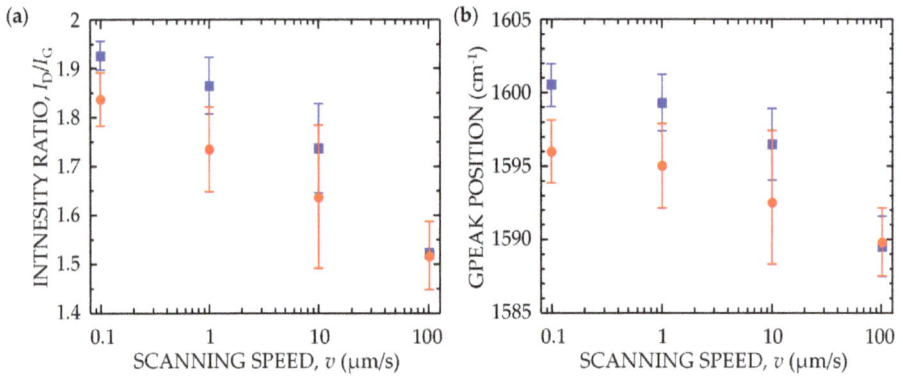

Figure 6. (a) Ratio of intensities of D and G peaks (I_D/I_G) and (b) position of G peak for DLC films irradiated with 7-fs laser pulses with $I = 1$ TW/cm^2 at $F = 6$ mJ/cm^2 (red circles) and those irradiated with 100-fs laser pulses with $I = 0.1$ TW/cm^2 at $F = 6$ mJ/cm^2 (blue squares) as function of scanning speed v.

Based on these experimental results and the physical mechanism for nanostructuring [8,15,16,30], the origin of the decrease in d with increasing F is discussed. The SPP wavelength λ_{spp} was calculated for the model surface illustrated in the inset of Figure 7, where it was assumed that the fs laser pulses are incident on the target in air, free electrons are produced at the GC surface to form a thin metal-like layer on the DLC substrate, and SPPs are excited at the interface between the metal-like layer and the DLC. The calculation method was almost the same as that used in our previous studies [15,16]. Briefly, $\lambda_{spp} = 2\pi/\text{Re}[k_{spp}]$ was calculated using the following relation between light and SPPs:

$$k_{spp} = k_0 \left[\varepsilon_{DLC} \, \varepsilon^*/(\varepsilon_{DLC} + \varepsilon^*)\right]^{1/2} \tag{1}$$

where k_0 is the wavevector of the incident light in vacuum, and ε^* and ε_{DLC} are the relative dielectric constants for the metallic GC and the DLC, respectively. As the GC layer is ionized by fs laser pulses, ε^* rapidly changes during the interaction as:

$$\varepsilon^* = \varepsilon_{GC} - [\omega_p^2/(\omega^2 + i\omega/\tau)] \tag{2}$$

where ε_{GC} is the static dielectric constant for the GC layer, and the second term represents the effect of free electrons with a density of N_e produced in the GC layer, where ω is the laser frequency in vacuum, $\tau = 1$ fs is the Drude damping time for free electrons [31,32], and $\omega_p = [e^2 N_e/(\varepsilon_0 \, m^* \, m)]^{1/2}$ is the plasma frequency, with the dielectric constant of vacuum ε_0, electron charge e, electron mass m, and optical effective mass of electrons $m^* = 1$. In the calculation, because the wavelength of the 7-fs laser pulse used in the present experiment was 680–940 nm, the static dielectric constants for DLC and GC were used for three wavelengths: $\varepsilon_{DLC} = 6.9 + i3.8$ and $\varepsilon_{GC} = 3.0 + i2.8$ for $\lambda = 600$ nm; $\varepsilon_{DLC} = 8.0 + i2.9$ and $\varepsilon_{GC} = 3.1 + i3.1$ for $\lambda = 800$ nm; and $\varepsilon_{DLC} = 8.5 + i2.6$ and $\varepsilon_{GC} = 3.6 + i4.5$ for $\lambda = 1000$ nm [33].

Figure 7 shows the period of the PNS, $D = \lambda_{spp}/2 = \pi/(Re[k_{spp}])$, calculated for $\lambda = 600, 800,$ and 1000 nm as a function of N_e. The excitation of SPPs at the interface between the metallic GC layer and the DLC is allowed for $Re[\varepsilon^*] \times Re[\varepsilon_{DLC}] < 0$ [20], which corresponds to the regions of $N_e > 1.0 \times 10^{22}$ cm^{-3} for $\lambda = 600$ nm, $N_e > 6.4 \times 10^{21}$ cm^{-3} for $\lambda = 800$ nm, and $N_e > 5.2 \times 10^{21}$ cm^{-3} for $\lambda = 1000$ nm. With increasing N_e, D decreases from ~200 nm to ~100 nm. Because N_e should increase with increasing I via stronger nonlinear optical absorption, the decrease in D with increasing N_e is in good agreement with the decrease in d with increasing I for 7-fs laser pulses shown in Figure 4.

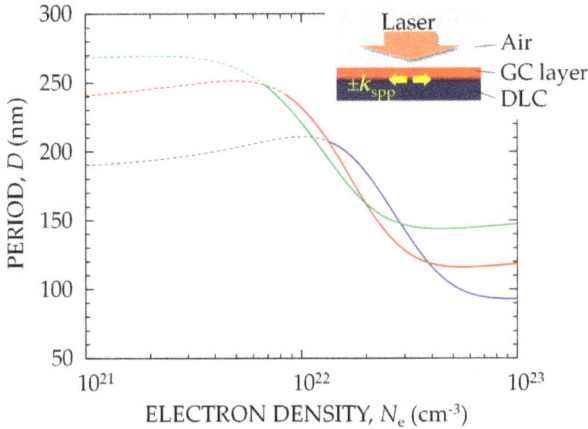

Figure 7. Calculated groove period D as function of N_e in the glassy carbon (GC) layer at $\lambda = 600$ nm (**blue**), 800 nm (**red**), and 1000 nm (**green**), where the excitation of surface plasmon polaritons (SPPs) is allowed in the region (**solid curves**) of $0 < Re[\varepsilon^*]$. The inset shows a schematic drawing of the initial target surface modeled for calculation. SPPs (**left/right arrows**) are excited at the interface between the DLC and GC layers by a high density of electrons produced by irradiation using high power -density laser pulses (**down arrow**).

The present experimental and calculation results show that the period d for a PNS was smaller than D, and that the d value for a PNS formed with high-fluence 100-fs laser pulses was similar to D for $\lambda = 800$ nm, which is consistent with the results of a previous study [15]. Regarding the excitation of SPPs on a thin metal film, it has been reported that the wavenumber of the SPPs increases with decreasing thickness of the film because of an increase in the radiation damping of SPPs [20]. These results suggest that d being smaller than D can be attributed to the excitation of SPPs with a larger wavenumber by the thinner metallic layer produced with 7-fs laser pulses. A calculation model for D that includes the effect of the metallic layer thickness will be presented and discussed in a separate paper. To discuss the formation process of PNS in detail and make a more accurate model for the nanostructuring, we need to quantitatively measure the amount and thickness of the GC layer on DLC film by using advanced techniques, such as a grazing-incidence small-angle X-ray scattering [34–36].

4. Conclusions

This study examined the PNS that formed on a DLC film with 7-fs laser pulses at a low fluence from a laser oscillator. The results show the formation of a PNS with a period of $d = 60$–80 nm and a decrease in d with increasing fluence. Based on the experimental results and a model calculation, it is shown that the excitation of SPPs at the interface between the thin metal-like layer and the DLC is certainly responsible for the nanostructuring process, and that the decrease of *d* is attributed to the wavelength of the SPPs decreasing with increasing *F* due to an increase of electron density in the thin metal-like layer.

Author Contributions: S.N. and R.S. performed the ablation experiments; T.N. fabricated the micro Raman spectroscope and analyzed the bonding structure of ablated surfaces; G.M. designed the experiments and made the physical model; all the authors wrote this paper together.

Funding: This research was supported in part by the Amada Foundation 2015 and Joint Usage/Research Program on Zero-Emission Energy Research, Institute of Advanced Energy, Kyoto University (ZE29C-1, 2017).

Acknowledgments: The authors would like to thank K.S. for the preliminary experiments, K.M. for helpful comments and useful discussions on Raman spectroscopy, and A.H. and Y.H. for the composition analysis of DLC films.

Conflicts of Interest: The authors declare no conflict of interest.

References and Note

1. Bonse, J.; Sturm, H.; Schmidt, D.; Kautek, W. Chemical, morphological and accumulation phenomena in ultrashort-pulse laser ablation of TiN in air. *Appl. Phys. A Mater. Sci. Process.* **2000**, *71*, 657–665. [CrossRef]
2. Reif, J.; Costache, F.; Henyk, M.; Pandelov, S.V. Ripples revisited: Non-classical morphology at the bottom of femtosecond laser ablation craters in transparent dielectrics. *Appl. Surf. Sci.* **2002**, *197–198*, 891–895. [CrossRef]
3. Yasumaru, N.; Miyazaki, K.; Kiuchi, J. Femtosecond-laser-induced nanostructure formed on hard thin films of TiN and DLC. *Appl. Phys. A Mater. Sci. Process.* **2003**, *76*, 983–985. [CrossRef]
4. Wu, Q.; Ma, Y.; Fang, R.; Liao, Y.; Yu, Q. Femtosecond laser-induced periodic surface structure on diamond film. *Appl. Phys. Lett.* **2003**, *82*, 1703–1705. [CrossRef]
5. Borowiec, A.; Haugen, H.K. Subwavelength ripple formation on the surfaces of compound semiconductors irradiated with femtosecond laser pulses. *Appl. Phys. Lett.* **2003**, *82*, 4462–4464. [CrossRef]
6. Daminelli, G.; Krüger, J.; Kautek, W. Femtosecond laser interaction with silicon under water confinement. *Thin Solid Films* **2004**, *467*, 334–341. [CrossRef]
7. Bonse, J.; Krüger, J. Femtosecond laser-induced periodic surface structures. *J. Appl. Phys.* **2012**, *24*, 042006-1–042006-7. [CrossRef]
8. Miyazaki, K.; Miyaji, G. Mechanism and control of periodic surface nanostructure formation with femtosecond laser pulses. *Appl. Phys. A Mater. Sci. Process.* **2014**, *114*, 177–185. [CrossRef]
9. Miyazaki, K.; Miyaji, G.; Inoue, T. Nanograting formation on metals in air with interfering femtosecond laser pulses. *Appl. Phys. Lett.* **2015**, *107*, 071103. [CrossRef]
10. Miyaji, G.; Miyazaki, K. Fabrication of 50-nm period gratings on GaN in air through plasmonic near-field ablation induced by ultraviolet femtosecond laser pulses. *Opt. Express* **2016**, *24*, 4648–4653. [CrossRef] [PubMed]
11. Yasumaru, N.; Miyazaki, K.; Kiuchi, J. Glassy carbon layer formed in diamond-like carbon films with femtosecond laser pulses. *Appl. Phys. A Mater. Sci. Process.* **2004**, *79*, 425–427. [CrossRef]
12. Miyazaki, K.; Maekawa, N.; Kobayashi, W.; Kaku, M.; Yasumaru, N.; Kiuchi, J. Reflectivity in femtosecond-laser-induced structural changes of diamond-like carbon film. *Appl. Phys. A Mater. Sci. Process.* **2005**, *80*, 17–21. [CrossRef]
13. Miyaji, G.; Miyazaki, K. Ultrafast dynamics of periodic nanostructure formation on diamond-like carbon films irradiated with femtosecond laser pulses. *Appl. Phys. Lett.* **2006**, *89*, 191902. [CrossRef]
14. Miyaji, G.; Miyazaki, K. Nanoscale ablation on patterned diamond-like carbon film with femtosecond laser pulses. *Appl. Phys. Lett.* **2007**, *91*, 123102. [CrossRef]
15. Miyaji, G.; Miyazaki, K. Origin of periodicity in nanostructuring on thin film surfaces ablated with femtosecond laser pulses. *Opt. Express* **2008**, *16*, 16265–16271. [CrossRef] [PubMed]
16. Miyaji, G.; Miyazaki, K.; Zhang, K.; Yoshifuji, T.; Fujita, J. Mechanism of femtosecond-laser-induced periodic nanostructure formation on crystalline silicon surface immersed in water. *Opt. Express* **2012**, *20*, 14848–14856. [CrossRef] [PubMed]

17. Miyazaki, K.; Miyaji, G. Nanograting formation through surface plasmon fields induced by femtosecond laser pulses. *J. Appl. Phys.* **2013**, *114*, 153108. [CrossRef]
18. Miyaji, G.; Miyazaki, K. Shaping of nanostructured surface in femtosecond laser ablation of thin films. *Appl. Phys. A Mater. Sci. Process.* **2010**, *98*, 927–930. [CrossRef]
19. Kärtner, F.X.; Morgner, U.; Schibli, T.; Ell, R.; Haus, H.A.; Fujimoto, J.G.; Ippen, E.P. Few-Cycle Pulses Directly from a Laser. In *Few-Cycle Laser Pulse Generation and Its Applications*, 1st ed.; Kärtner, F.X., Ed.; Springer: Berlin/Heidelberg, Germany, 2004; pp. 73–136. ISBN 978-3-540-20115-1.
20. Raether, H. *Surface Plasmons on Smooth and Rough Surfaces and on Gratings*; Springer-Verlag: Heidelberg, Germany, 1988; ISBN 978-3-540-47441-8.
21. Yoshikawa, M.; Katagiri, G.; Ishida, H.; Ishitani, A. Raman spectra of diamondlike amorphous carbon films. *J. Appl. Phys.* **1988**, *64*, 6464–6468. [CrossRef]
22. Zhang, S.; Zeng, X.T.; Xie, H.; Hing, P. A phenomenological approach for the Id/Ig ratio and sp3 fraction of magnetron sputtered a-C films. *Surf. Coat. Technol.* **2000**, *123*, 256–260. [CrossRef]
23. Robertson, J. Diamond-like amorphous carbon. *Mater. Sci. Eng.* **2002**, *R37*, 129–281. [CrossRef]
24. Tai, F.C.; Lee, S.C.; Wei, C.H.; Tyan, S.L. Correlation between I_D/I_G Ratio from Visible Raman Spectra and sp^2/sp^3 Ratio from XPS Spectra of Annealed Hydrogenated DLC Film. *Mater. Trans.* **2006**, *47*, 1847–1852. [CrossRef]
25. Tuinstra, F.; Koenig, J.L. Raman Spectrum of Graphite. *J. Chem. Phys.* **1970**, *53*, 1126–1130. [CrossRef]
26. Nemanich, R.J.; Solin, S.A. First- and second-order Raman scattering from finite-size crystals of graphite. *Phys. Rev. B* **1979**, *20*, 392–400. [CrossRef]
27. Yoshikawa, M.; Nagai, N.; Matsuki, M.; Fukuda, H.; Katagiri, G.; Ishida, H.; Ishitani, A.; Nagai, I. Raman scattering from sp^2 carbon clusters. *Phys. Rev. B* **1992**, *46*, 7169–7174. [CrossRef]
28. Ferrari, A.C.; Robertson, J. Interpretation of Raman spectra of disordered and amorphous carbon. *Phys. Rev. B* **2000**, *61*, 14095–14107. [CrossRef]
29. Wojdyr, M. Fityk: A general-purpose peak fitting program. *J. Appl. Crystal.* **2010**, *43*, 1126–1128. [CrossRef]
30. Miyaji, G.; Miyazaki, K. Role of multiple shots of femtosecond laser pulses in periodic surface nanoablation. *Appl. Phys. Lett.* **2013**, *103*, 071910. [CrossRef]
31. Sokolowski-Tinten, K.; Linde, D. Generation of dense electron-hole plasmas in silicon. *Phys. Rev. B* **2000**, *61*, 2643–2650. [CrossRef]
32. Because the Drude damping time of the DLC films or GC was never measured, we used the value of Si reported in Ref. 31. Using the time in a range of 1–10 fs, we confirmed that the calculation results were almost the same as that with the time of 1 fs.
33. Alterovitz, S.A.; Savvides, N.; Smith, F.W.; Woollam, J.A. Amorphous Hydrogenated "Diamondlike" Carbon Films and Arc-Evaporated Carbon Films. In *Handbook of Optical Constants of Solids*; Palik, E.D., Ed.; Academic Press: San Diego, CA, USA, 1985; pp. 838–852. ISBN 978-0-125-44423-1.
34. Rebollar, E.; Pérez, S.; Hernández, J.J.; Martín-Fabiani, I.; Rueda, D.R.; Ezquerra, T.A.; Castillejo, M. Assessment and Formation Mechanism of Laser-Induced Periodic Surface Structures on Polymer Spin-Coated Films in Real and Reciprocal Space. *Langmuir* **2011**, *27*, 5596–5606. [CrossRef] [PubMed]
35. Rebollar, E.; Rueda, D.R.; Martín-Fabiani, I.; Rodríguez-Rodríguez, Á.; García-Gutiérrez, M.-C.; Portale, G.; Castillejo, M.; Ezquerra, T.A. In Situ Monitoring of Laser-Induced Periodic Surface Structures Formation on Polymer Films by Grazing Incidence Small-Angle X-ray Scattering. *Langmuir* **2015**, *31*, 3973–3981. [CrossRef] [PubMed]
36. Roth, S.V.; Döhrmann, R.; Gehrke, R.; Röhlsberger, R.; Schlage, K.; Metwalli, E.; Körstgens, V.; Burghammer, M.; Riekel, C.; David, C.; et al. Mapping the morphological changes of deposited gold nanoparticles across an imprinted groove. *J. Appl. Crystallogr.* **2015**, *48*, 1827–1833. [CrossRef]

nanomaterials

MDPI

Article

Formation of Slantwise Surface Ripples by Femtosecond Laser Irradiation

Xin Zheng [1,2], Cong Cong [3], Yuhao Lei [1], Jianjun Yang [1,*] and Chunlei Guo [1,4,*]

[1]   The Guo China-US Photonics Laboratory, State Key Laboratory of Applied Optics,
    Changchun Institute of Optics, Fine Mechanics and Physics, Chinese Academy of Sciences,
    Changchun 130033, China; lindsax@163.com (X.Z.); leiyuhao93@gmail.com (Y.L.)
[2]   College of Materials Science and Opto-Electronic Technology, University of Chinese Academy of Science,
    Beijing 100049, China
[3]   School of the Gifted Young, University of Science and Technology of China, Hefei 230026, China;
    cong2014@mail.ustc.edu.cn
[4]   The Institute of Optics, University of Rochester, Rochester, NY 14627, USA
*   Correspondence: jjyang@ciomp.ac.cn (J.Y.); guo@optics.rochester.edu (C.G.);
    Tel.: +86-138-2070-1265 (J.Y.); +1-585-275-2134 (C.G.)

Received: 26 May 2018; Accepted: 19 June 2018; Published: 22 June 2018

Abstract: We report on the formation of slantwise-oriented periodic subwavelength ripple structures on chromium surfaces irradiated by single-beam femtosecond laser pulses at normal incidence. Unexpectedly, the ripples slanted in opposite directions on each side the laser-scanned area, neither perpendicular nor parallel to the laser polarization. The modulation depth was also found to change from one ripple to the next ripple. A theoretical model is provided to explain our observations, and excellent agreement is shown between the simulations and the experimental results. Moreover, the validity of our theory is also confirmed on bulk chromium surfaces. Our study provides insights for better understanding and control of femtosecond laser nanostructuring.

Keywords: femtosecond laser; laser-induced periodic surface structures; anomalous slanting ripples; chromium

---

1. Introduction

During the last several years, the research of femtosecond laser-induced periodic surface structures (Fs-LIPSSs), or the ripple structures, has attracted tremendous attention because of the abundant scientific issues involved [1–3]. Fs-LIPSSs have been studied on a variety of materials, including metals, semiconductors, and dielectrics [4–7]. It has been found that such microstructures have extensive potential applications, such as magnetic recording media [8], self-cleaning materials [9,10], anti-reflective metals [11], and solar sensors [12].

In general, the distinct characteristics of the ripple structures are closely dependent on the laser parameters. When the linearly polarized single-beam femtosecond laser pulses are used to irradiate materials, the induced ripple structures are either parallel or perpendicular to the direction of the laser polarization [13–16]. In some cases, however, the ripple structures induced by femtosecond lasers presented an unusual feature of slantwise orientation, which is neither perpendicular nor parallel to the laser polarization direction [17–22]. For instance, Qiu et al. [17] reported the slantwise oriented nanogrooves on a ZnO crystal surface with normal incidence of the single-beam femtosecond laser, which tended to be perpendicular to the direction of the laser polarization at the increased scanning speed. By tilting the incident angle of femtosecond laser pulses, Schwarz et al. [19] experimentally observed the slantwise orientation of the ripple structures on fused silica, and the structure orientation changed as a function of the laser incident angle. More recently, our research group generated a series of v-like structures, called a herringbone pattern, on copper [23]. Such anomalous phenomena indicate the physical complexities during the ripple surface structure formation, which is actually significant for femtosecond laser nanoprocessing. Nevertheless, a comprehensive underlying mechanisms of LIPSS orientation is still lacking.

In this work, the formation of slantwise oriented ripple structures is systematically investigated on chromium surfaces by employing single-beam femtosecond laser pulses at normal incidence. First, the ripple structures generated on two lateral edges of the laser-scanned area are seen to have different slantwise orientations with respect to the direction of the laser polarization, and such behaviors occur even when the laser polarization changes. Secondly, based on the measured modulation depth of the ripple structures, we develop a theoretical model to elucidate the underlying mechanisms via the consistent simulations. Finally, additional experiments are performed to confirm the theory.

2. Materials and Methods

As shown by a schematic illustration of the experimental setup in Figure 1, a commercial chirped-pulse-amplification of a Ti:sapphire laser system (Spitfire Ace, Spectra Physics, Santa Clara, CA, USA) was employed as a light source for producing the surface structures, which delivers horizontally polarized femtosecond laser pulse trains with a repetition rate, a central wavelength, and a time duration of 1 kHz, 800 nm, and 40 fs, respectively. The maximum energy of each laser pulse was 7 mJ. Neutral density filters and a half-wave plate were used to control the pulse energy and the direction of the laser polarization, respectively. The laser beam was focused by an objective lens (plan fluorite objective, 4×, N.A = 0.13, f = 17.2 mm, Nikon, Tokyo, Japan) at normal incidence. The sample was mounted on a three-dimensional translation stage (ESP301, Newport Inc., Irvine, CA, USA) that could be precisely translated via a custom-made computer program. In order to avoid serious ablation of the material, the sample surface was moved 300 μm away from the focus towards the lens, such that the focus is located inside the sample, resulting in a laser spot radius of ≈39 μm on the sample surface.

In the experiments, 100 nm-thickness chromium (Cr) films deposited on SiO_2 substrate were chosen as sample materials because of its good physical characteristics, including hardness, corrosion resistance, high melting point, and adhesiveness, which earn many applications in solar absorbers, adhesion layers, micro-electromechanical systems devices, etc. [24–26]. Besides, we also used bulk Cr material to carry out experiments. Based on the method of the previous reports [25,27], we experimentally obtained the ablation threshold values of a single laser pulse for the film and bulk of Cr material, which can be given as about 37.5 mJ/cm^2 and 248 mJ/cm^2, respectively. All the experimental performances were carried out in ambient air environment. After the laser irradiation, the surface morphologies were investigated using a scanning electron microscope (SEM, Phenom, Eindhoven, Netherlands) and an atomic force microscope (AFM, Bruker, Billerica, MA, USA).

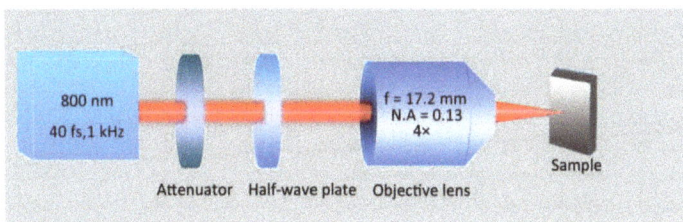

Figure 1. A schematic experimental setup for the formation of slantwise-oriented ripple structures on chromium surface by femtosecond laser irradiation.

3. Results and Discussions

Figure 2a exhibits the surface morphology of the Cr film irradiated by single-beam femtosecond laser pulses at the fluence of $F = 56.9$ mJ/cm^2 with the scanning speed of $V = 0.3$ mm/s. Due to the incident Gaussian laser intensity, different surface morphologies could be observed on the area where the laser scanned; substantial ablation damages occurred in the central area, and there were ripple structures on the lateral edges, with a spatial period of approximate 650 nm. In particular, the periodic ripple structures on both edge regions were found to have a slantwise orientation in two different directions. This behavior is in sharp contrast to the previous reports [13–16], wherein the ripple orientation is usually either perpendicular or parallel to the direction of the incident laser polarization.

To further characterize the features of such slantwise-oriented ripple structures, we employed AFM to measure their modulation depths, and the corresponding results are shown in Figure 2b. Clearly, the measured oscillation curve reveals that the modulation depth of the ripple structures decreases gradually with increasing the distance from the center to the lateral edges of the laser-scanned area, which is due to the spatially inhomogeneous distribution of the laser pulse intensity. On the other hand, the measured peaks suggest that the modulation height of the surface is also varied as a function of the distance from the center of the laser-scanned area. This can be physically understood as follows: The film thickness decreased after irradiation of multiple femtosecond laser pulses, leading to the film thinning at the center of the laser-scanned area with respect to the lateral edge regions. Consequently, the formation of such slantwise oriented periodic ripple structures is in fact based on the gradient variation of the film thicknesses.

Noticeably, the measured height of the ripple structures was larger than the thickness of the Cr film, which may have been due to material reaction with O$_2$ in the ambient atmosphere, leading to oxide formation on the material surface [28]. More specifically, as shown in Figure 2c, there are two physical processes happening in the formation of the laser-induced ripple structures: one is the spatially periodic removal of chromium materials by the modulated laser intensity fringes, and the other is the growth of chromium oxides at the places where the laser intensity is higher than the threshold of oxidation. Usually, the laser damage threshold is larger than that of the oxidation process. Here it should be clear that the periodic femtosecond laser intensity distribution for the ripple structure formation is originated from interference of the light and its excited surface plasmons [29–32]. Because the two components possess unequal energies, i.e., the energy of the excited surface plasmons is usually smaller than that of the incident laser pulse, their interfering intensity patterns, which had a Gaussian variation profile tend to give a low fringe contrast, or the deconstructive interference fringes can also hold a certain level of the laser energy. Under such circumstances, the material oxidation can take place during the formation of the periodic ripple structures, and the resultant additional oxide layers on the ridge surfaces make the height of the ripple structures become protuberant with respect to the original film thickness.

Inspired by the anomalous phenomenon of the ripple structures with the slantwise orientation, we also performed a series of experiments on Cr films by varying the direction of linear polarization of the femtosecond laser. As shown by the results in Figure 3 (here only the observations on both lateral edges of the laser-scanned area are shown), for the given laser polarization, the slantwise-oriented periodic ripple structures are always produced on both lateral edge regions of the laser-scanned area, being very similar to the observation in Figure 2. Whereas for different laser polarizations, the slantwise degree of the ripple structures is found to change but still neither perpendicular nor parallel to the laser polarization direction. Therefore, we can conclude that the formation of slantwise orientated periodic ripple structures seems to always appear even for different linear polarizations of femtosecond laser pulses.

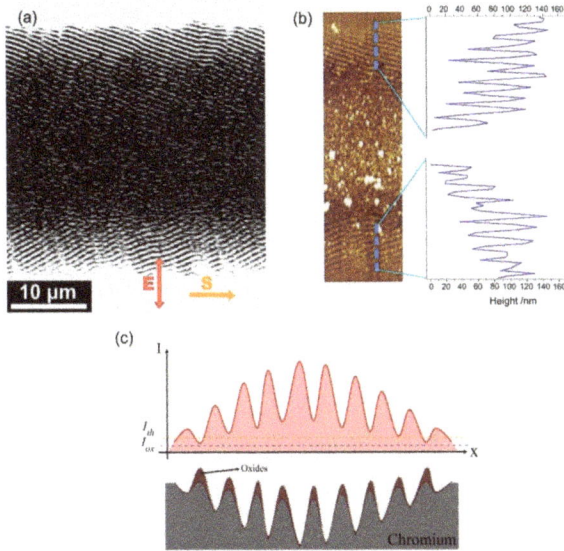

Figure 2. (a) SEM image of the ripple structure formation on a Cr film surface irradiated by single-beam femtosecond laser irradiation at the energy fluence of $F = 56.9$ mJ/cm^2 with the sample scanning speed of $V = 0.3$ mm/s; (b) AFM image with the cross-section profiles of the ripple structures formed on both lateral edge regions of the laser-scanned area. Arrows of **S** and **E** represent directions of the sample scanning and the laser polarization, respectively; (c) Schematic plots of the periodically distributed intensity distribution on the Cr surface (upper), and its induced ripple structures (bottom), where I_{th} and I_{ox} indicate the threshold intensities of the material damage and oxidation processes, respectively. The oxidation layer on the top parts of the surface structures are represented by a purple color.

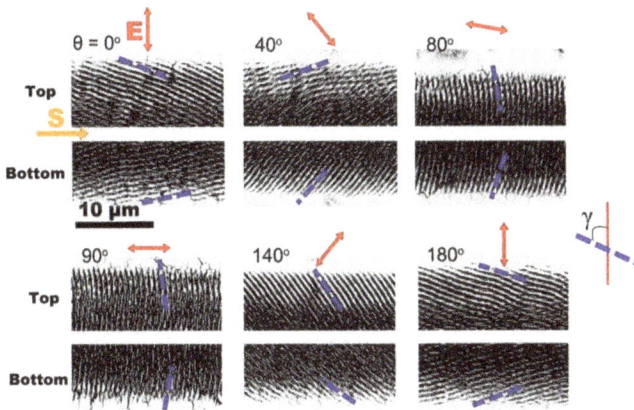

Figure 3. Slantwise-oriented ripple structures on two lateral edge regions (Top and Bottom) of the laser-scanned area on Cr film surfaces by different linear polarizations of single-beam femtosecond laser pulses. The angle θ on the upper-left corner of each image represents the direction of the laser polarization. The blue dash lines identify the orientations of the ripple structures. The angle of γ indicates an intersection angle between the ripple orientation and the laser polarization direction of $\theta = 0°$. The scale bar is applied to all images in this figure.

To elucidate our experimental observations, we proposed the following physical scenario: In our experiments, which are in fact based on multi-pulse femtosecond laser irradiation processes, the pristine surface of the metal film was modified by the preceding incident femtosecond laser pulses, leading to a rough, shallow crater with the modulation depth gradually reducing from the beam center to the peripheral regimes, as shown in Figure 4a, where the inclined surface was created on the laser irradiation area. After that, for the continuous irradiation of the subsequent femtosecond laser pulses, the inclining degree of the laser irradiation surface became pronounced (Figure 4b), and the periodic subwavelength ripple structures were also developed on it, exhibiting the slantwise orientation with respect to the direction of the laser polarization, as shown in Figure 4c. Noticeably, due to the higher intensity distribution on the central region of the laser-scanned area, the formation of the corresponding ripple structures was seriously deteriorated by the accumulating irradiation of subsequent femtosecond laser pulses.

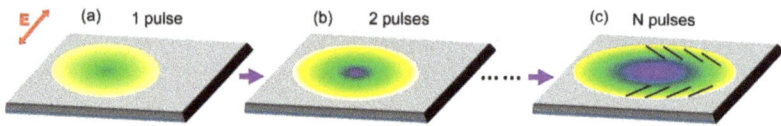

Figure 4. Schematic diagrams of the physical processes for the formation of slantwise-oriented periodic ripple structures on the metal surface. **E** represents the direction of the laser polarization. The different colors represent variations of the modulation depth, which tended to cause the inclined surface within the laser irradiation area.

In fact, the effects of the inclined surface on the formation of slantwise-orientated periodic ripple structures can be theoretically analyzed. According to the previous study of Pham et al. [33], the presence of the inclined surface on the laser spot area can be described by $pX + qY + Z = 1$, within a three-dimensional Cartesian coordinate system X-Y-Z, as shown in Figure 5a. Here p and q are the geometrical parameters for describing the spatial characteristics of the inclined surface. A normal component of the inclined surface is represented by $\mathbf{n} = (p, q, 1)$. Both the propagation direction and the electric field vectors of the incident femtosecond laser are defined as $\mathbf{L_i} = (0, 0, 1)$ and $\mathbf{E_i} = (\cos\theta, \sin\theta, 0)$, respectively, wherein θ is an intersection angle between $\mathbf{E_i}$ and the X-axis. As shown in Figure 5a, a plane of the laser incidence (represented by a blue color) is established by the vectors of $\mathbf{n}$ and $\mathbf{L_i}$, whose intersection angle is defined by θ_i. Moreover, a coordinate system x'-y'-z' is also built for simplifying the calculation of the electric field on the inclined surface. For the incidence of femtosecond laser on the inclined surface, its electric field vector $\mathbf{E_i}$ is divided into two components of $\mathbf{E_{x'}}$ and $\mathbf{E_m}$ through its projection onto the x'-y' and the incident planes, respectively. On the other hand, the projection of the electric field component $\mathbf{E_m}$ on the x'-y' plane is indicated by $\mathbf{E_{y'}}$. Finally, in the x'-y' plane, the two electric field components $\mathbf{E_{x'}}$ and $\mathbf{E_{y'}}$ can be developed into a new vector of $\mathbf{E_{x'y'}}$, as shown in Figure 5b, with β being an intersection angle between $\mathbf{E_{x'}}$ and $\mathbf{E_{x'y'}}$, which is calculated by the following expression [33]:

$$\beta = \arctan\left(\sqrt{\frac{\left(n\cos\theta_i\sqrt{n^2 - \sin^2\theta_i} + n^3 - n\sin^2\theta_i\right)^2 + \kappa^2\left(n^2 - \sin^2\theta_i\right)^2}{\left(n\sqrt{n^2 - \sin^2\theta_i} + n^3\cos\theta_i\right)^2 + n^4\kappa^2\cos^2\theta_i}}\cot\alpha\right) \tag{1}$$

where $\theta_i = \arccos\dfrac{1}{\sqrt{1+p^2+q^2}}$ and $\alpha = \arcsin\dfrac{q\cos\theta - p\sin\theta}{\sqrt{p^2+q^2}}$.

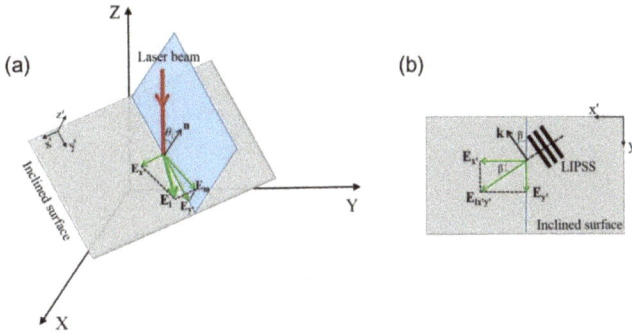

Figure 5. (**a**) A sketch of the laser incidence onto the inclined surface of the material and the decomposition of the electric field $\mathbf{E_i}$ onto different planes; (**b**) An effective electric field vector on the x'-y' plane and its induced periodic ripple structures with the orientation vector $\mathbf{k}$.

In Equation (1), n and κ represent the real and the imaginary parts of the complex refractive index $\tilde{n}$ of the material, respectively. Accordingly, the orientation vector of the ripple structures on the inclined surface, $\mathbf{k}$, should be perpendicular to the direction of the electric field $\mathbf{E_{x'y'}}$, as shown in Figure 5b. When the ripple orientation $\mathbf{k}$ in the x'-y'-z' coordinate system is transferred into the X-Y-Z coordinate system, it should be modified into:

$$\mathbf{k} = \left(q\sin\beta - \frac{p\cos\beta}{\sqrt{1+p^2+q^2}}, \ -p\sin\beta - \frac{q\cos\beta}{\sqrt{1+p^2+q^2}}, \ \frac{(p^2+q^2)\cos\beta}{\sqrt{1+p^2+q^2}} \right) \qquad (2)$$

By considering the actual observation surface happening on the X-Y plane, the orientation vector $\mathbf{k}$ can be re-written as:

$$\mathbf{k} = \left(q\sin\beta - \frac{p\cos\beta}{\sqrt{1+p^2+q^2}}, \ -p\sin\beta - \frac{q\cos\beta}{\sqrt{1+p^2+q^2}}, \ 0 \right) \qquad (3)$$

In the experiments, the orientation vector $\mathbf{k}$ was obtained by the measurement of angle γ shown in Figure 3. Thus, the assumed geometric parameters (p, q) of the inclined surface could be calculated by the non-linear fitting of Equation (3) with the help of the measured values $\mathbf{k} = (\cos\gamma, \sin\gamma, 0)$. For example, with the experimentally measured angles of γ, the achieved p and q values were $(p_{top} = -0.6399, q_{top} = -0.8141)$ and $(p_{bottom} = -0.5077, q_{bottom} = 0.6690)$ for the top and bottom edges of the laser-scanned area, respectively.

Through combing the above calculated p and q values with the expression of $pX + qY + Z = 1$, we could map three-dimensional profiles of the two inclined surfaces, as shown in Figure 6a, where the left and right surfaces indicate the top and bottom edges of the laser-scanned area, respectively. Evidently, for each inclined surface, the modulation height is varied as a function of the x-y position. In addition, we could also calculate the ripple orientation for the single beam femtosecond laser at different polarization directions. Specifically, because the structure orientation is indicated by the vector $\mathbf{k} = (\cos\gamma, \sin\gamma, 0)$ where $\cos\gamma = q\sin\beta - \frac{p\cos\beta}{\sqrt{1+p^2+q^2}}$ and $\sin\gamma = -p\sin\beta - \frac{q\cos\beta}{\sqrt{1+p^2+q^2}}$, we could obtain γ values with the available parameters of p and q. Therefore, by changing the laser polarization from $\theta = 0°$ to $180°$, the theoretical fitting of the ripple orientation angle γ on the top and bottom edges of the laser-scanned area could be obtained, as shown (by red solid curves) in Figure 6b,c, respectively, wherein the experimental data are given (by blue solid circles) with the standard deviation. It is seen clearly that the theory and experiment have good consistency in the two cases. Another feature is that the obtained ripple orientation angle vs the laser polarization direction had nonlinear variations, which was basically due to the polarization dependent optical absorption.

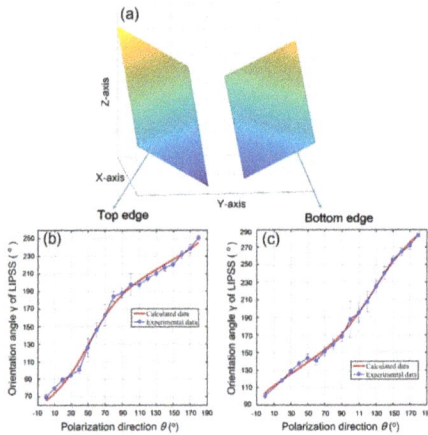

Figure 6. (a) Theoretically retrieving the inclined surfaces on the top and bottom edges of the laser-scanned area in the coordinate system of X-Y-Z, with the help of the calculated parameters of (p_{top} = −0.6399, q_{top} = −0.8141) and (p_{bottom} = −0.5077, q_{bottom} = 0.6690), respectively; (b,c) compare the simulation results with the experimental data for the ripple orientation angles on the top and bottom edges of the laser-scanned area, respectively.

In order to confirm the above theoretical analyses, we carried out further experiments on the surface of Cr bulk material. As shown in Figure 7a, when the femtosecond laser energy fluence was given at F = 40.6 mJ/cm^2, the subwavelength ripple structures with orientation perpendicular to the laser polarization can still be formed in the central parts of the laser-scanned area. While on both lateral edges (marked by the red dot frames) of the laser-scanned area, the obtained ripple structures exhibit different slantwise orientations. From the corresponding AFM image, as shown in Figure 7b, we can also find that the modulation of the ripple depth is decreased with larger distances from the center of the laser scribed area, which is attributed to the Gaussian beam profile distribution. Clearly, the measured varying tendencies of the ripple depth on the Cr bulk material are very similar to the observations on Cr films.

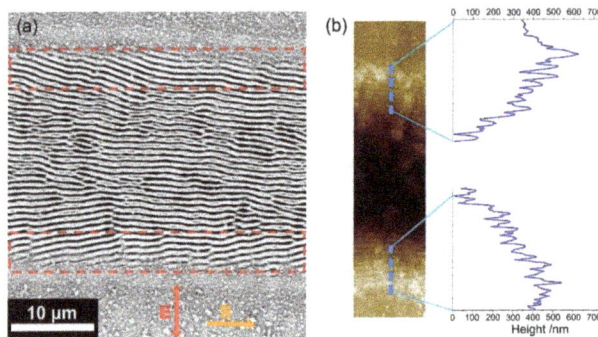

Figure 7. (a) SEM image of the ripple structure formation on the surface of Cr bulk material irradiated by single-beam femtosecond laser pulses at the energy fluence of F = 40.6 mJ/cm^2; (b) AFM image with the cross-section profiles for the ripple structures on two edge regions of the laser-scanned area.

On the other hand, when the femtosecond laser energy fluence was decreased to approximately F = 25.7 mJ/cm^2, the ripple structures turned out to be oriented perpendicular to the direction of the laser polarization, especially on both lateral edges of the laser-scanned area, shown in Figure 8a, being similar to many previous reports [13–16]. As a matter of fact, this situation can be maintained even for the laser energy fluences of about $F = 29.6$ mJ/cm^2, as shown in Figure 8b, where the ripple-covered region seemed to be enlarged with the orientation still perpendicular to the direction of the laser polarization. Based on the experimental comparisons, it is revealed that the spatial alignment of the ripple structures can be transferred from the slantwise tendency into the direction perpendicular to the laser polarization, if the femtosecond laser energy fluence is weak enough. In other words, the incident larger energy fluence of the Gaussian laser beam was a key factor for the formation of the inclined surface during the multi-pulse laser irradiation, which finally resulted in the slantwise oriented ripple structures on the lateral edges of the laser-scanned area.

Such a ripple orientation-transferring process can be understood as follows: For a Gaussian laser pulse irradiation on the material with a damage threshold intensity of I_{th}, the resultant ablation depth varied as a function of the distance away from the beam center. Therefore, the gradient surfaces were likely to be modified on the ablation edges, with an incline angle φ proportional to the variation rate of the laser intensity, i.e., $\varphi \propto \sqrt{\ln \frac{I_0}{I_{th}}}$, where I_0 was the peak intensity of the laser pulse. Evidently, with increasing the laser energy fluence, the higher peak intensity could result in a larger incline degree of the ablation surface, as shown in Figure 8c, and the consequent formation of the slantwise oriented ripple structures. Whereas for a femtosecond laser with lower energy fluences, the peak intensity caused a smaller incline degree of the ablation surface, providing negligible influence on orientation of the ripple structures.

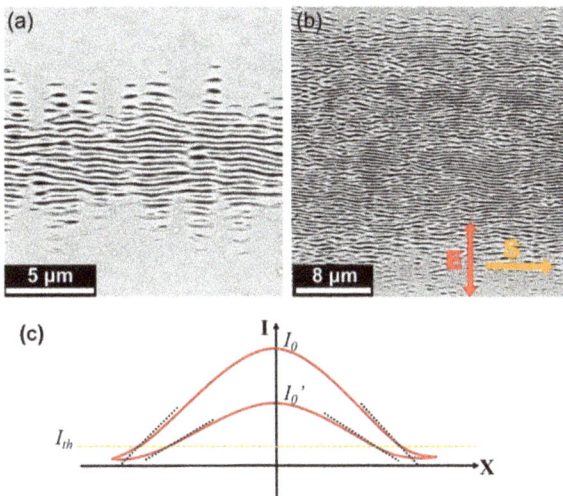

Figure 8. SEM images of the Cr bulk surfaces irradiated by single-beam femtosecond laser pulses with the different energy fluences. (**a**) $F = 25.7$ mJ/cm^2; (**b**) $F = 29.6$ mJ/cm^2; (**c**) Variation rates (dot curves) of the laser intensity at the damage threshold (I_{th}) for two cases of different pulse energy fluences, where I_0 and $I_{0'}$ are the peak intensities of two laser pulses.

4. Conclusions

In conclusion, we comprehensively studied the generation of slantwise-oriented subwavelength periodic ripple structures on the surfaces of chromium material by normal incidence of linearly polarized femtosecond laser pulses. Our experimental results on chromium films demonstrated that the ripple structures formed on two lateral edges of the laser-scanned area are slantwise-oriented in two different directions, being neither perpendicular nor parallel to the laser polarization. When the laser polarization direction is changed, the slantwise ripple structures were still observable but with different orientations, AFM measurements suggested that the modulation height of the ripple-covered surface exhibited gradual variations with the distance from the center to the lateral edges, which is due to the inhomogeneous distribution of the Gaussian laser intensity. A physical model was proposed by considering the inclined ablation surfaces after multi-pulse irradiations. The agreement between the simulations and the measured results confirmed the validity of our theory.

The above mentioned slantwise ripple structures were also generated on bulk Cr surfaces. For the reduced laser fluence, however, the slantwise observations began to transfer into the commonly observed ripple structures with an orientation perpendicular to the direction of the laser polarization, which indicated the negligible influence of the inclined ablation surface. Our investigation provides a comprehensive understanding of femtosecond laser-material interactions, which may help us design and fabricate uniform subwavelength and even nanoscale structures and devices for the future applications.

Author Contributions: C.G. and J.Y. conceived and designed the experiments; X.Z. and Y.L. performed the experiments; C.C. performed the simulations; X.Z., Y. L., and C.C. analyzed the data; X.Z. and J.Y. wrote the paper.

Funding: The authors would like to acknowledge the support from National Key R&D Program of China (2017YFB1104700); National Natural Science Foundation of China (11674178, 61774155,); Natural Science Foundation of Tianjin City (17JCZDJC37900); Jilin Provincial Science & Technology Development Project (20180414019GH).

Conflicts of Interest: The authors declare no conflict of interest.

References

1. Krüger, J.; Kautek, W. The femtosecond pulse laser: A new tool for micromachining. *Laser Phys.* **1999**, *9*, 30–40.
2. Vorobyev, A.Y.; Guo, C.L. Direct femtosecond laser surface nano/microstructuring and its applications. *Laser Photonics Rev.* **2013**, *7*, 385–407. [CrossRef]
3. Sugioka, K.; Cheng, Y. Ultrafast lasers—Reliable tools for advanced materials processing. *Light Sci. Appl.* **2014**, *3*, e149. [CrossRef]
4. Wang, J.C.; Guo, C.L. Ultrafast dynamics of femtosecond laser-induced periodic surface pattern formation on metals. *Appl. Phys. Lett.* **2005**, *87*, 251914. [CrossRef]
5. Xue, L.; Yang, J.J.; Yang, Y.; Wang, Y.; Zhu, X. Creation of periodic subwavelength ripples on tungsten surface by ultra-short laser pulses. *Appl. Phys. A* **2012**, *109*, 357–365. [CrossRef]
6. Borowiec, A.; Haugen, H.K. Subwavelength ripple formation on the surfaces of compound semiconductors irradiated with femtosecond laser pulses. *Appl. Phys. Lett.* **2003**, *82*, 4462–4464. [CrossRef]
7. Hnatovsky, C.; Taylor, R.S.; Rajeev, P.P.; Simova, E.; Bhardwaj, V.R.; Rayner, D.M.; Corkum, P.B. Pulse duration dependence of femtosecond-laser-fabricated nanogratings in fused silica. *Appl. Phys. Lett.* **2005**, *87*, 014104. [CrossRef]
8. Stupakiewicz, A.; Szerenos, K.; Afanasiev, D.; Kirilyuk, A.; Kimel, A.V. Ultrafast nonthermal photo-magnetic recording in a transparent medium. *Nature* **2017**, *542*, 71. [CrossRef] [PubMed]
9. Wu, B.; Zhou, M.; Li, J.; Ye, X.; Li, G.; Cai, L. Superhydrophobic surfaces fabricated by microstructuring of stainless steel using a femtosecond laser. *Appl. Surf. Sci.* **2009**, *256*, 61–66. [CrossRef]
10. Vorobyev, A.Y.; Guo, C.L. Multifunctional surfaces produced by femtosecond laser pulses. *J. Appl. Phys.* **2015**, *117*, 033103. [CrossRef]
11. Vorobyev, A.Y.; Makin, V.S.; Guo, C.L. Brighter light sources from black metal: Significant increase in emission efficiency of incandescent light sources. *Phys. Rev. Lett.* **2009**, *102*, 234301. [CrossRef] [PubMed]
12. Hwang, T.Y.; Vorobyev, A.Y.; Guo, C.L. Enhanced efficiency of solar-driven thermoelectric generator with femtosecond laser-textured metals. *Opt. Express* **2011**, *19* (Suppl. 4), A824–A829. [CrossRef] [PubMed]
13. Qi, L.T.; Nishii, K.; Namba, Y. Regular subwavelength surface structures induced by femtosecond laser pulses on stainless steel. *Opt. Lett.* **2009**, *34*, 1846–1848. [CrossRef] [PubMed]

14. He, W.L.; Yang, J.J.; Guo, C.L. Controlling periodic ripple microstructure formation on 4h-sic crystal with three time-delayed femtosecond laser beams of different linear polarizations. *Opt. Express* **2017**, *25*, 5156–5168. [CrossRef] [PubMed]
15. Jia, X.; Jia, T.Q.; Zhang, Y.; Xiong, P.X.; Feng, D.H.; Sun, H.Y.; Qiu, J.R.; Xu, Z.Z. Periodic nanoripples in the surface and subsurface layers in zno irradiated by femtosecond laser pulses. *Opt. Lett.* **2010**, *35*, 1248–1250. [CrossRef] [PubMed]
16. Tang, Y.F.; Yang, J.J.; Zhao, B.; Wang, M.W.; Zhu, X.N. Control of periodic ripples growth on metals by femtosecond laser ellipticity. *Opt. Express* **2012**, *20*, 25826–25833. [CrossRef] [PubMed]
17. Liu, J.K.; Jia, T.Q.; Zhou, K.; Feng, D.H.; Zhang, S.A.; Zhang, H.X.; Jia, X.; Sun, Z.R.; Qiu, J.R. Direct writing of 150 nm gratings and squares on zno crystal in water by using 800 nm femtosecond laser. *Opt. Express* **2014**, *22*, 32361–32370. [CrossRef] [PubMed]
18. Yang, Y.; Yang, J.J.; Liang, C.; Wang, H.; Zhu, X.; Kuang, D.; Yang, Y. Sub-wavelength surface structuring of niti alloy by femtosecond laser pulses. *Appl. Phys. A* **2008**, *92*, 635–642. [CrossRef]
19. Schwarz, S.; Rung, S.; Hellmann, R. One-dimensional low spatial frequency lipss with rotating orientation on fused silica. *Appl. Surf. Sci.* **2017**, *411*, 113–116. [CrossRef]
20. Jiang, L.; Shi, X.S.; Li, X.; Yuan, Y.P.; Wang, C.; Lu, Y.F. Subwavelength ripples adjustment based on electron dynamics control by using shaped ultrafast laser pulse trains. *Opt. Express* **2012**, *20*, 21505–21511. [CrossRef] [PubMed]
21. Pan, A.; Dias, A.; Gomez-Aranzadi, M.; Olaizola, S.M.; Rodriguez, A. Formation of laser-induced periodic surface structures on niobium by femtosecond laser irradiation. *J. Appl. Phys.* **2014**, *115*, 173101. [CrossRef]
22. Petrović, S.M.; Gaković, B.; Peruško, D.; Stratakis, E.; Bogdanović-Radović, I.; Čekada, M.; Fotakis, C.; Jelenković, B. Femtosecond laser-induced periodic surface structure on the ti-based nanolayered thin films. *J. Appl. Phys.* **2013**, *114*, 233108. [CrossRef]
23. Garcell, E.M.; Lam, B.; Guo, C. Femtosecond laser-induced herringbone patterns. *Appl. Phys. A* **2018**, *124*, 405. [CrossRef]
24. Kotsedi, L.; Nuru, Z.Y.; Mthunzi, P.; Muller, T.F.G.; Eaton, S.M.; Julies, B.; Manikandan, E.; Ramponi, R.; Maaza, M. Femtosecond laser surface structuring and oxidation of chromium thin coatings: Black chromium. *Appl. Surf. Sci.* **2014**, *321*, 560–565. [CrossRef]
25. Banerjee, S.P.; Fedosejevs, R. Single-shot ablation threshold of chromium using UV femtosecond laser pulses. *Appl. Phys. A* **2014**, *117*, 1473–1478. [CrossRef]
26. Sue, J.A.; Chang, T.P. Friction and wear behavior of titanium nitride, zirconium nitride and chromium nitride coatings at elevated temperatures. *Surf. Coat. Technol.* **1995**, *76–77*, 61–69. [CrossRef]
27. Saghebfar, M.; Tehrani, M.K.; Darbani, S.M.R.; Majd, A.E. Femtosecond pulse laser ablation of chromium: Experimental results and two-temperature model simulations. *Appl. Phys. A* **2017**, *123*, 28. [CrossRef]
28. Öktem, B.; Pavlov, I.; Ilday, S.; Kalaycıoğlu, H.; Rybak, A.; Yavaş, S.; Erdoğan, M.; Ilday, F.Ö. Nonlinear laser lithography for indefinitely large-area nanostructuring with femtosecond pulses. *Nat. Photonics* **2013**, *7*, 897–901. [CrossRef]
29. Huang, M.; Zhao, F.L.; Cheng, Y.; Xu, N.S.; Xu, Z.Z. Origin of laser-induced near-subwavelength ripples: Interference between surface plasmons and incident laser. *ACS Nano* **2009**, *3*, 4062–4070. [CrossRef] [PubMed]
30. Sakabe, S.; Hashida, M.; Tokita, S.; Namba, S.; Okamuro, K. Mechanism for self-formation of periodic grating structures on a metal surface by a femtosecond laser pulse. *Phys. Rev. B* **2009**, *79*, 033409. [CrossRef]
31. Garrelie, F.; Colombier, J.P.; Pigeon, F.; Tonchev, S.; Faure, N.; Bounhalli, M.; Reynaud, S.; Parriaux, O. Evidence of surface plasmon resonance in ultrafast laser-induced ripples. *Opt. Express* **2011**, *19*, 9035–9043. [CrossRef] [PubMed]
32. Okamuro, K.; Hashida, M.; Miyasaka, Y.; Ikuta, Y.; Tokita, S.; Sakabe, S. Laser fluence dependence of periodic grating structures formed on metal surfaces under femtosecond laser pulse irradiation. *Phys. Rev. B* **2010**, *82*, 165417. [CrossRef]
33. Pham, K.X.; Tanabe, R.; Ito, Y. Laser-induced periodic surface structures formed on the sidewalls of microholes trepanned by a femtosecond laser. *Appl. Phys. A* **2012**, *112*, 485–493. [CrossRef]

nanomaterials

MDPI

Article

Subwavelength Nanostructuring of Gold Films by Apertureless Scanning Probe Lithography Assisted by a Femtosecond Fiber Laser Oscillator

Ignacio Falcón Casas * and Wolfgang Kautek *

Department of Physical Chemistry, University of Vienna, Währinger Strasse 42, Vienna A-1090, Austria
* Correspondence: ignacio.falcon.casas@univie.ac.at (I.F.C.); wolfgang.kautek@univie.ac.at (W.K.);
   Tel.: +43-664-8175230 (W.K.)

Received: 5 June 2018; Accepted: 11 July 2018; Published: 16 July 2018

Abstract: Optical methods in nanolithography have been traditionally limited by Abbe's diffraction limit. One method able to overcome this barrier is apertureless scanning probe lithography assisted by laser. This technique has demonstrated surface nanostructuring below the diffraction limit. In this study, we demonstrate how a femtosecond Yb-doped fiber laser oscillator running at high repetition rate of 46 MHz and a pulse duration of 150 fs can serve as the laser source for near-field nanolithography. Subwavelength features were generated on the surface of gold films down to a linewidth of 10 nm. The near-field enhancement in this apertureless scanning probe lithography setup could be determined experimentally for the first time. Simulations were in good agreement with the experiments. This result supports near-field tip-enhancement as the major physical mechanisms responsible for the nanostructuring.

Keywords: near-field; femtosecond laser; nanolithography; subwavelength; tip-enhancement; AFM

---

1. Introduction

Optical methods in lithography are limited by Abbe's diffraction limit. Apertureless scanning probe nanolithography assisted by a laser is a method able to overcome this limitation [1–9]. In this technique, a sharp scanning probe microscope (SPM) tip is placed a few nanometres above the surface of a substrate. The tip is irradiated by a laser and a strong enhanced field may be generated in the proximity of the apex of the tip. The evanescent near-field decays exponentially in both lateral and vertical axes, which leads to a confinement of the electromagnetic field. This may surpass the modification fluence threshold of the substrate. The combination of tip-enhancement and confinement can be employed to produce sub-wavelength surface nanostructuring. However, although near-field enhancement has been predicted in simulations [10–13] and observed experimentally [14–20], a number of unresolved issues still exist. An example is how much thermal effects are contributing. On one hand, a laser-irradiated tip can reach a high temperature [21–23], leading to the destruction or modification of the tip itself. Even if the tip is not modified, heat transfer from the hot tip to the substratee—either by conduction or radiation—may lead to the melting of the substrate. Recent research has found huge near-field heat transfer coefficient values, whose origin has not been clarified yet [24–26]. On the other hand, laser irradiation can produce a fast bending and expansion of cantilevers due to thermal diffusion, leading to mechanical indentation or scratching of the surface of the substrate [23]. This problem can be handled by using low spring constant cantilevers [4] or keeping the tip in non-contact mode [8].

Femtosecond lasers have demonstrated excellent performance in material ablation because the heat affected zone can be minimized to a few nanometres and can avoid laser–plasma interactions completely. Continuous-wave lasers have also been employed in scanning probe optical lithography, but the use of ultrashort pulsed lasers allows for reducing the average power applied to the tip, while producing very high intensities. Although these high intensities can be beneficial regarding substrate structuring, laser intensities above the MW/cm^2 level might lead to tip damage. Special care has to be taken with metal-coated tips, as it has been reported in apertureless scanning near-field microscopy (aSNOM) experiments [27].

Apertureless scanning probe near-field lithography assisted by femtosecond laser has demonstrated subwavelength surface structuring of metal and polymer films, reaching linewidths down to 10–15 nm [4,8]. These experiments were performed with Ti:Sa laser sources, at laser wavelengths $\lambda \approx 790$ nm. Theoretical calculations have shown that, under some conditions, longer laser wavelengths might lead to higher near-field enhancement [28,29]. A combination of high repetition rate and low energy per pulse helps to reduce thermomechanical instabilities of the tip-cantilever [30].

In the present study, apertureless scanning probe near-field lithography was employed to write nanofeatures on the surface of gold films. A home-built femtosecond Ytterbium-doped fiber laser oscillator based on a novel design [31] was applied. A repetition rate of 46 MHz and a low pulse energy was chosen to allow a better thermomechanical equilibrium of the cantilever in contrast to laser sources with kHz repetition rates and higher energy per pulse [30]. The generated subwavelength grooves exhibited typical linewidths of 40–60 nm, down to 10 nm, thus surpassing the diffraction limit. The near-field enhancement could be determined experimentally for the first time, and was compared with simulations.

2. Materials and Methods

We employed a scanning probe microscope NTEGRA (NT-MDT, Moscow, Russia) working in atomic force microscope (AFM) contact mode to scan the surface of the samples. AFM measurements were done in ambient conditions. A closed-loop configuration of the scanner stage was enabled to increase the lateral positioning accuracy. We used silicon probes with a radius of curvature of 10 nm (CSG10, NT-MDT) at the apex of the tip. Contact mode low spring constant cantilevers were chosen to avoid mechanical deformation of the gold films. The measured resonance frequency of the cantilever was 27 KHz, 30 μm width and 225 μm length. By using these values, and following the Sader's method [32], we obtained a spring constant of the cantilever k = 0.4 N/m. We determined the load force applied against the sample from the spring constant value of the cantilever. We performed force spectroscopy on a silicon calibration grating (TGZ1, NT-MDT, Moscow, Russia) in order to obtain a linear relationship between cantilever deflection (in nA) and cantilever height (in nm). A deflection setpoint value of 2 nA was chosen when scanning in contact mode, corresponding to a load force of $F_{load} \approx 20$ nN. This low value of the load force was chosen to avoid the possibility of mechanical surface modification. The scanning probe microscope data were evaluated with the software Gwyddion.

Gold films were prepared by thermal evaporation. Two films were deposited on mica (15 and 30 nm thickness) and another on a glass substrate (30 nm thickness). Details of AFM surface characterization of the samples are shown in Appendix B.

A scheme of the experimental setup is shown in Figure 1. The laser source for the experiments was a home-built femtosecond Ytterbium-doped fiber laser oscillator, designed according to [31]. The output of the laser cavity was directed to a pair of diffraction gratings that compensate the dispersion and compress the laser pulse. The laser beam consisted of laser pulses with temporal length $\tau = 150$ fs, 1 nJ energy per pulse, central wavelength $\lambda = 1040$ nm, 46 MHz repetition rate, and linear polarization. The polarization was controlled by a half-wave plate. The laser power was adjusted by a combination of another half-wave plate and a polarizing beamsplitter cube. After passing through a periscope, the laser beam was focused onto the SPM tip by an achromatic lens (AC254-060-B-ML, Thorlabs, Newton, NJ, USA) with a focal length f = 60 mm. The laser spot diameter at the focal plane was 50 μm and the angle of incidence $\theta = 88°$ with respect to the tip axis. Under this condition, the tip with a height of ca. 15 μm was entirely irradiated. The laser peak intensity was adjusted during the experiments from a minimum of $I = 0.15 \times 10^8$ W/cm^2 to a maximum value of $I = 2.90 \times 10^8$ W/cm^2.

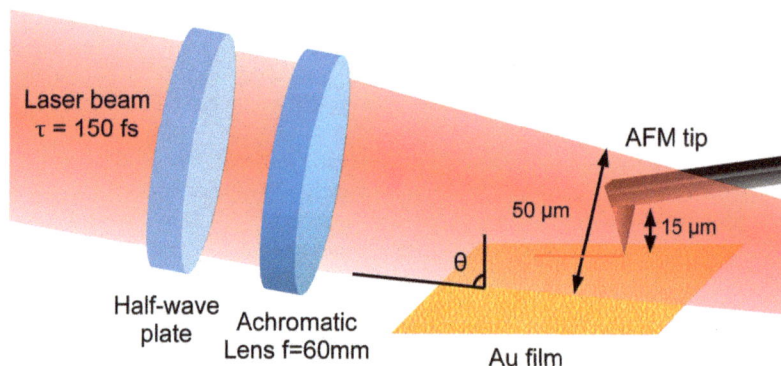

Figure 1. A scheme of the experimental setup. The p-polarized laser beam is focused onto an atomic force microscope (AFM) tip by an achromatic lens. The tip height is about 15 μm, so the laser focal spot size (50 μm) illuminates the tip entirely. A half-wave plate allows for controlling the laser polarization angle.

The nanostructuring procedure consisted of three steps. First, an AFM scan of the sample surface was performed to obtain the topography before laser irradiation. The stage scanned one line along the fast axis (y-axis) and then moved along the slow axis (x-axis) to the next line position, where the scan along the fast axis was repeated. In the second step, we scanned the same area and, at a certain position of the slow scanning axis, the scanning in the slow direction (x-axis) was paused. Then, the laser beam was engaged and the tip was irradiated, while moving along the fast axis (y-axis). After a number of laser pulses, the laser beam was turned off and the scan continued. Finally, we scan again the same area to observe any change on the surface.

Near-field simulations were performed using a MNPBEM17 toolbox, which is based on a boundary element method (BEM) [33,34]. We introduced the values of the refractive indexes of silicon at 300 K [35] and 25 nm thick gold films [36]. The simulations were performed taking into account the retardation of the electromagnetic fields, by solving the full Maxwell equations.

3. Results

3.1. Near-Field Enhancement Simulations

Near-field enhancement simulations were performed for a gold substrate and two silicon tip geometries, a sphere and a rod (Figure 2). The enhancement factor γ can be defined as $\gamma = |E|/|E_0|$, where E is the induced electric field on the tip-substrate and E_0 is the initial laser electric field. A p-polarized electromagnetic plane wave irradiated the tip at an angle of incidence $\theta = 88°$ (referred to the axis normal to the substrate). The wavelength of the incoming laser light was set at $\lambda = 1040$ nm, which corresponds to the central wavelength of our fiber laser. We obtained near-field enhancement factors of $\gamma \approx 5$–15 for the sphere, depending on the sphere's radius of curvature and the tip-substrate distance (Figure 2a). The enhancement factor of the rod was higher than for the sphere and changed with the length L of the rod. Typical values ranged from $\gamma = 12$ (L = 70 nm, r = 10 nm) to $\gamma = 220$ (L = 200 nm, r = 10 nm) (Figure 2b). In both cases, the maximum field enhancement was located at the particle-substrate interspace and decreased exponentially with distance.

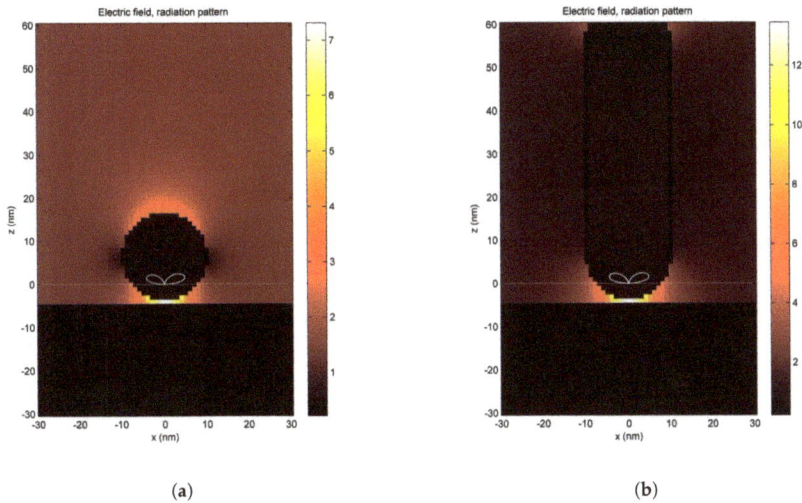

Figure 2. Near-field enhancement between a gold nanofilm and: (**a**) a spherical particle (radius = 10 nm); (**b**) a rod (r = 10 nm, length = 70 nm) both placed 1 nm above a gold surface. A p-polarized electromagnetic plane wave irradiates at λ = 1040 nm, with an angle of incidence θ = 88°. The white lines represent the direction and magnitude of the Poynting vector of far-field scattered laser light.

3.2. Scanning Probe Near-Field Nanolithography

In this section, we present the results of the laser irradiation and nanostructuring of gold nanofilms. In order to identify the irradiation parameters in the far-field which lead to irreversible modifications of the nanofilms, the threshold fluence as a function of the angle of incidence θ and the number of pulses N were determined (without the AFM tip engaged).

The influence of the number of pulses N was studied on a 15 nm thick gold film on mica. An AFM image of the surface of the sample before laser irradiation can be seen on Figure 3a. The gold surface is composed of nanocrystalline islands with typical sizes of about 50–100 nm, similar to gold film surfaces reported in [37]. Far-field laser irradiation was performed at an angle of incidence of θ = 86° and an intensity of $I = 2.4 \times 10^8$ W/cm^2. The scanning direction was set from left to right, at a scanning speed of 1 μm/s. A substantial morphological modification was observed after an irradiation time of 90 seconds, corresponding to a number of pulses $N = 4.2 \times 10^9$ (Figure 3b). There, the gold film is strongly deformed and forms bumps that elevate to heights up to 100 nm. This experiment served to identify morphological surface changes produced by far-field irradiation.

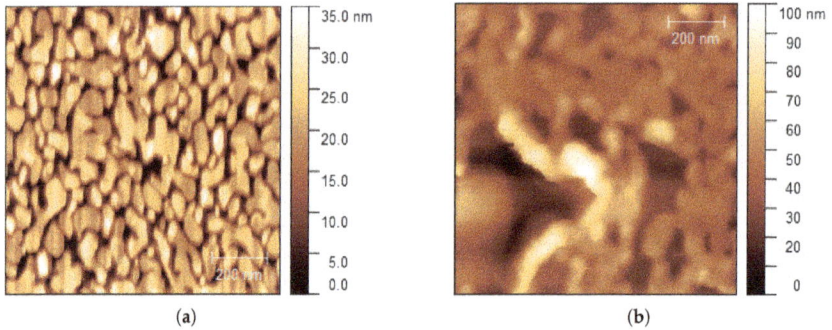

(a)   (b)

Figure 3. Morphology changes produced by the far-field laser. Gold on mica (15 nm thickness) (a) before laser irradiation; (b) after a laser irradiation time of 90 seconds, corresponding to $N = 4.2 \times 10^9$. Angle of incidence $\theta = 86°$, $I = 2.4 \times 10^8$ W/cm². Scan area 1×1 µm².

To analyze the influence of the angle of incidence, a similar experiment was conducted on a 30 nm thick gold film on glass at $\theta = 80°$. Far-field surface modifications similar to Figure 3b were observed (not shown here), even for a lower number of pulses $N = 0.19 \times 10^9$. Therefore, we inferred that the angle of incidence has a drastic effect on far-field surface modifications, more significant than the number of pulses. In order to produce near-field nanolithography, far-field surface modifications need to be avoided. Therefore, the laser intensity was decreased by changing the angle of incidence to $\theta = 88°$. At this angle of incidence, no far-field surface modifications were observed, even for a high number of pulses $N = 4.2 \times 10^9$. Figure 4a shows the surface of the sample (30 nm thick gold film on glass) before laser irradiation. Similar gold grain patterns can be observed before and after laser irradiation (e.g., green boxes in Figure 4a,b). This indicates that the far-field laser did not affect the surface morphology and only the areas below the tip were modified during irradiation.

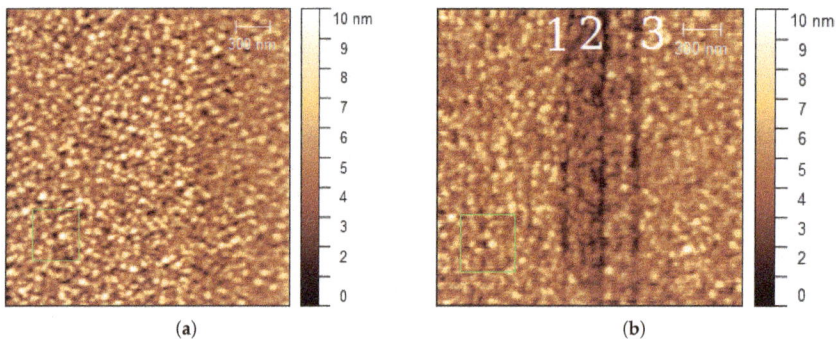

(a)   (b)

Figure 4. Nanolithography dependence on the number of scans at tip illumination. Gold on glass (30 nm thickness) (a) before laser irradiation and (b) after laser irradiation at three lines (slow axis scanning from left to right stopped). Number of scans: 7 (line 1), 10 (line 2) and 14 (line 3), scanning speed 0.62 µm/s, angle of incidence $\theta = 88°$, $I = 2.4 \times 10^8$ W/cm².

Figure 4b shows the effect produced by repeated tip passes on the same line under laser illumination. The tip was irradiated at three lines (slow axis scanning stopped) during 30, 60 and 90 s ($N = 1.2, 2.4$ and 4.2×10^9, respectively) (from left to right), corresponding to a number of tip passes on each line of 7, 10 and 14 times, respectively. Three vertical lines were structured on the gold surface (Figure 4b). The linewidth and depth both increased with the number of passes (Table 1).

Table 1. Dependence of width (FWHM – Full Width at Half Maximum) and depth of lines on the number of passes (in Figure 4b, from left to right). Data are averaged from 256 profile lines.

	Line 1	Line 2	Line 3
Number of passes	7	10	14
Width (nm)	36	45	76
Depth (nm)	0.4	0.5	1.0

The dependence on the laser intensity was investigated on a 30 nm thick gold film on mica (Figure 5a). The tip was irradiated at four vertical lines (slow axis scanning from left to right stopped) at increasing laser intensities $I = 0.2, 0.3, 0.4, 1.0 \times 10^8$ W/cm^2 (from left to right). The scanning speed was set at 0.38 µm/s. Laser intensities below 0.4×10^8 W/cm^2 produced a very small effect. The depth of lines increased with the laser intensity. A line with an averaged FWHM width of 10 nm (Figure 5b) was produced at $I = 1.0 \times 10^8$ W/cm^2. The line profile was obtained by averaging the green rectangle area marked in Figure 5a.

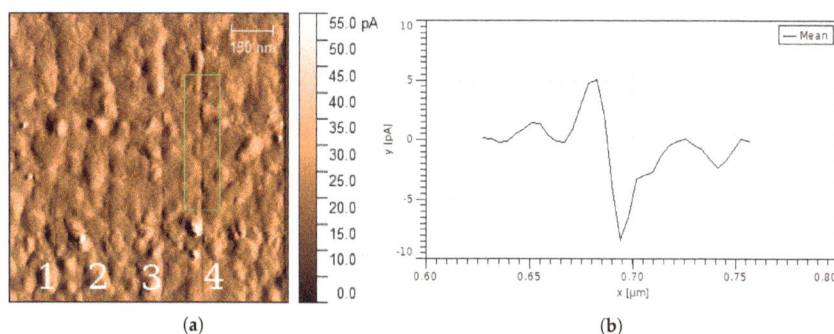

(a)   (b)

Figure 5. Nanolithography depending on laser intensity. Gold on mica (30 nm thickness) (a) after laser irradiation at four vertical lines (slow axis scanning from left to right stopped). $I = 0.2, 0.3, 0.4, 1.0 \times 10^8$ W/cm^2, angle of incidence $\theta = 88°$, scanning speed 0.38 µm/s; (b) averaged line profile of the area marked in Figure 5a. The FWHM width of the line is 10 nm. The image was obtained in AFM contact error mode, where a constant force is applied to the tip and the variations of the cantilever's deflection are recorded.

Laser intensity dependence was also studied on a 30 nm thick gold film on glass. Figure 6a shows five vertical lines irradiated at increasing laser power $I = 0.7, 1.0, 2.0, 2.7$ and 2.9×10^8 W/cm^2 (from left to right) and scanning speed of 0.38 µm/s. The profile line was averaged (taking the full area in Figure 6a), to reduce the effect of the gold roughness. The FWHM width and depth of lines (Figure 6b) increased with the laser intensity (Table 2). The writing performance at lines 1, 4 and 5 was affected by 1–3 nm height irregularities of the gold surface.

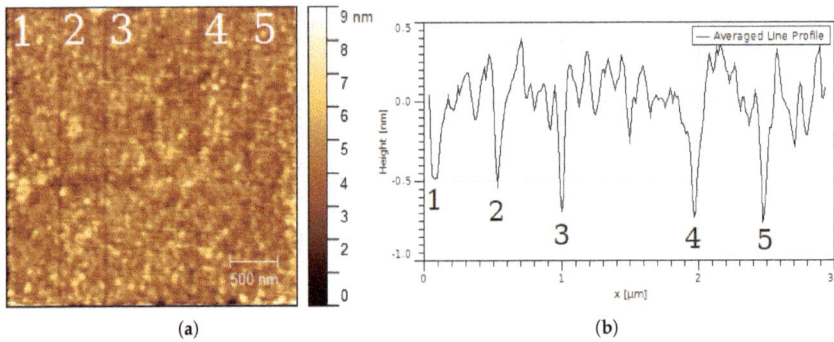

(a)   (b)

Figure 6. Nanolithography depending on the laser intensity. Gold on glass (30 nm thickness (**a**) after laser irradiation at five lines (slow axis scanning from left to right stopped). Laser intensity $I = 0.7$ (line 1), 1.0 (line 2), 2.0 (line 3), 2.7 (line 4) and 2.9×10^8 W/cm^2, scanning speed 0.38 μm/s, angle of incidence $\theta = 88°$; (**b**) line profile obtained after taking an average of 256 horizontal profile lines. Widths and depths are indicated in Table 2.

Table 2. Laser intensity effect on width (FWHM) and depth of lines in Figure 6b (from left to right). Data are averaged from 256 profile lines.

Line Number	1	2	3	4	5
I ($\times 10^8$W/cm^2)	0.7	1.0	2.0	2.7	2.9
Width (nm)	71	47	41	60	52
Depth (nm)	0.48	0.52	0.70	0.74	0.77

A summary of the laser far-field parameters is provided in Table 3.

Table 3. Far-field laser irradiation parameters for gold film on mica (15 nm thick) and on glass (30 nm thick): angle of incidence θ, laser peak intensity I, pulse fluence F, number of pulses N.

θ	I ($\times 10^8$W/cm^2)	F (μJ/cm^2)	N ($\times 10^9$)
80°	2.4	37	0.2
86°	2.4	37	1.4–4.2
88°	0.15–2.9	2–44	1.4–4.2

4. Discussion

Nanostructures were generated on the surface of gold films by using apertureless scanning near-field lithography. The structured lines formed on the gold films show typical linewidths of 40–70 nm and depths of 0.4–1.0 nm. The width of lines increased with the number of repeated tip scans and laser intensity. The smallest linewidth measured was 10 nm. The surface roughness of the gold samples affected the structuring performance (for instance, fourth and fifth lines in Figure 5). A threshold for surface modification at $I \approx 0.4 \times 10^8$ W/cm^2 (Figure 5a) was observed at a scanning speed of 0.38 μm/s. The angle of incidence (i.e., the laser intensity absorbed by the gold film) has a drastic effect on the far-field irradiation, as it can be seen in Figure 3b. At high angles of incidence (near 90°), the contrast between the near-field fluence near the tip and the far-field fluence on the illuminated substrate is sufficiently high so only nanostructures are generated without modification of the substrate (Figure 4b).

A comparatively low threshold fluence of $F_{th} \approx 6$ μJ/cm^2 for surface modification on gold was observed. This is in contrast to experiments with moderate repetition rates (1 kHz, $F_{th} = 12$ mJ/cm^2) [4], with single pulses (F_{th} = 2 mJ/cm^2) [38,39] and calculations with single pulses (F_{th} = 4.5 mJ/cm^2) [40]. Near-field surface modification was also achieved at high repetition rates but lower energy per pulse (e.g., 80 MHz, 1 nJ) [8]. This seems to indicate that the repetition rate has a strong influence on the modification of the gold surface. The so-called *cool-ablation* [41] has been observed at very high repetition rates of GHz using bursts of laser pulses. Threshold fluences can be reduced when the repetition rate is increased above 1–10 MHz [41,42].

Two main mechanisms have been proposed in the context of apertureless scanning probe near-field lithography assisted by laser: near-field tip-enhancement and thermomechanical effects. Simulations in Figure 2 show the presence of an enhanced electromagnetic field between the tip and the substrate. The major question is whether this enhancement factor is high enough to raise the laser field above the modification threshold of the substrate. An experimental estimation of the near-field enhancement, based on the observation of far-field morphological surface changes of the gold films, was undertaken. In the experiments shown in the Figure 3, we observed evidence of complete melting of the gold film after a high number of pulses N = 4.2 × 10^9 (Figure 3b). Based on this observation, the enhancement factor γ was determined by varying the angle of incidence on the 30 nm thick gold films. At an angle of incidence θ = 80°, the surface morphology was modified (not shown here), similarly to Figure 3b. Accordingly, the far-field fluence was above the modification threshold. An increase to an angle of incidence θ = 88° resulted in a reduction of the absorbed laser energy by a factor of $I_{88°}/I_{80°} \approx 0.02$ (see Appendix A). Under these conditions, one observes near-field modification without far-field modification (see Figure 4). In a first approximation, one could conclude that the enhancement is at least $I_{80°}/I_{88°} \approx 50$.

This result is now compared with a simulated enhancement factor. For the spherical tip shown in Figure 2a, the electric field enhancement is $\gamma \approx 7$. The intensity enhancement is $\gamma^2 = |E|^2/|E_0|^2 = 49$, in good agreement with the intensity threshold obtained experimentally ($I_{80°}/I_{88°} \approx 50$).

5. Conclusions

Nanostructuring of gold films by apertureless scanning probe near-field lithography using a novel Ytterbium-doped fiber oscillator as the laser source was demonstrated. Lines written on gold show depths of 0.5–1.0 nm and typical lateral sizes of 40–70 nm, down to 10 nm. A near-field enhancement factor was determined experimentally and compared with simulations. A good agreement with the experimental results supports a mechanism based mainly on near-field enhancement. This approach provides direct access to the study of femtosecond laser-matter interaction in the near-field nano-regime. The compact setup, the high repetition rate, the possibility of working in ambient conditions and comparatively reduced cost make this an appealing technique for sub-100 nm lithography.

Author Contributions: Conceptualization, I.F.C and W.K.; Formal analysis, I.F.C; Investigation, I.F.C; Resources, W.K.; Software, I.F.C; Supervision, W.K.; Visualization, I.F.C; Writing – original draft, I.F.C; Writing – review and editing, I.F.C and W.K.

Funding: This research received no external funding

Acknowledgments: Open access funding was provided by the University of Vienna. We thank Stefan Hummel for the preparation of one of the gold films.

Conflicts of Interest: The authors declare no conflict of interest.

Appendix A. Laser Absorbance Change versus Angle of Incidence for a Gold Nanofilm

The laser intensity threshold for surface modification was estimated by calculating the ratio between the absorbed laser intensity at two angles of incidence: θ = 80° (far-field laser irradiation produced morphology changes of the gold surface) and 88° (no far-field changes, only near-field when the tip was engaged). The two differences considered are the variation of both intensity and absorption with the incident angle. The first one can be calculated through the relationship

$$I(\theta) = I_0 cos^2(\theta), \tag{A1}$$

obtaining

$$\frac{I(80°)}{I(88°)} \approx 25. \tag{A2}$$

Then, the change in the absorbed laser intensity by the gold surface at the two different angles of incidence can be calculated. By using the refractive index at $\lambda = 1040$ nm for 25 nm thick gold films [36] and the Fresnel equations for p-polarization, one can calculate the transmittance T

$$r_p = \frac{n_2 cos\theta_i - n_1 cos\theta_t}{n_1 cos\theta_t + n_2 cos\theta_i}, \tag{A3}$$

$$T = 1 - R = 1 - |r_p|^2, \tag{A4}$$

where n_1, n_2 are the refractive indexes of air and gold, θ_i, θ_t are the incident and transmission angles, r_p is the Fresnel reflection coefficient for p-polarization and R is the reflectivity. For simplicity, we assume that the transmitted laser light is completely absorbed by the gold substrate. This assumption is a good approximation because the penetration depth of light in gold ($\delta \approx 12$ nm at normal incidence) is less than the gold film thickness. The transmittance values obtained at the angles of incidence considered are

$$\frac{T(80°)}{T(88°)} \approx 2. \tag{A5}$$

A total absorbed intensity ratio is obtained by the multiplication of the ratios of Equations (A2) and (A5). This means that, when the angle of incidence is changed from $\theta = 80°$ to $\theta = 88°$, the absorbed laser intensity is reduced by a factor of 50.

Appendix B. Morphology Characterization of the Surface of Au Films

The characterization of the topography of the Au films by AFM is presented here. Constant force and deflection error modes were recorded simultaneously by the AFM software (Nova, NT-MDT). Posterior analysis and calculation of average roughness values R_a were performed with the software Gwyddion. The average roughness R_a is defined as

$$R_a = \frac{1}{n} \sum_{i=1}^{n} |h_i|, \tag{A6}$$

where h is the height of each pixel and n is the number of pixels. In addition, 256×256 pixels (full images) were included for the calculations of R_a.

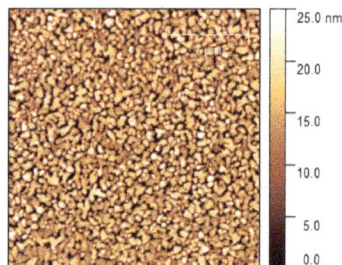

Figure A1. Topography of the 15 nm thickness Au film on mica imaged in AFM constant force mode; average surface roughness $R_a = 4.31$ nm, scan size 3×3 μm^2.

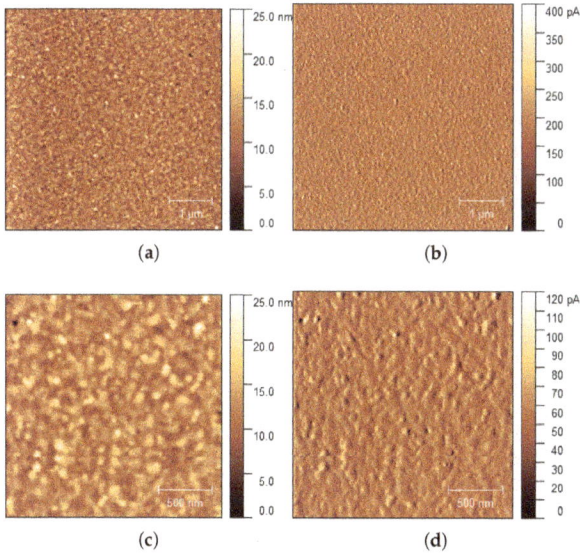

Figure A2. Topography of the 30 nm thickness Au film on mica imaged in AFM constant force mode (**a,c**) and error deflection mode (**b,d**). Average surface roughness R_a = 1.71 nm (**a**), 20.84 pA (**b**), 1.75 nm (**c**) and 7.55 pA (**d**), scan size 5 × 5 μm^2 (**a,b**) and 2 × 2 μm^2 (**c,d**).

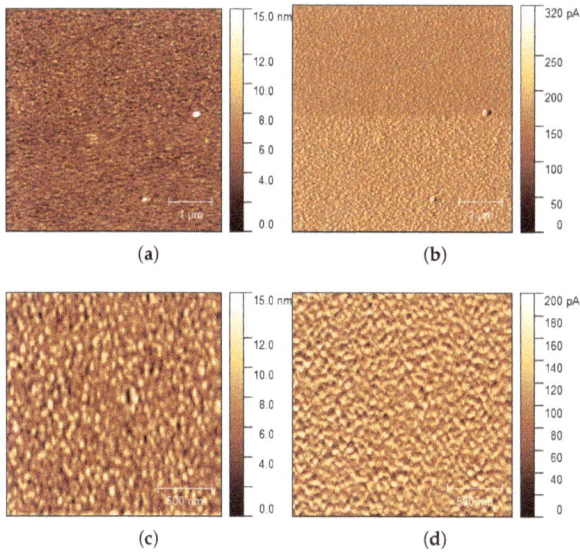

Figure A3. Topography of the 30 nm thickness Au film on glass imaged in AFM constant force mode (**a,c**) and error deflection mode (**b,d**). Average surface roughness R_a = 1.12 nm (**a**), 18.88 pA (**b**), 1.53 nm (**c**) and 21.45 pA (**d**), scan size 5 × 5 μm^2 (**a,b**) and 2 × 2 μm^2 (**c,d**).

References

1. Jersch, J.; Demming, F.; Dickmann, K. Nanostructuring with laser radiation in the nearfield of a tip from a scanning force microscope. *Appl. Phys. A* **1996**, *64*, 29–32. [CrossRef]

2. Dickmann, K.; Jersch, J.; Demming, F. Focusing of Laser Radiation in the Near-field of a Tip (FOLANT) for Applications in Nanostructuring. *Surf. Interface Anal.* **1997**, *25*, 500–504. [CrossRef]

3. Yin, X.; Fang, N.; Zhang, X.; Martini, I.B.; Schwartz, B.J. Near-field two-photon nanolithography using an apertureless optical probe. *Appl. Phys. Lett.* **2002**, *81*, 3663–3665. [CrossRef]

4. Chimmalgi, A.; Choi, T.Y.; Grigoropoulos, C.P.; Komvopoulos, K. Femtosecond laser apertureless near-field nanomachining of metals assisted by scanning probe microscopy. *Appl. Phys. Lett.* **2003**, *82*, 1146–1148. [CrossRef]

5. H'dhili, F.; Bachelot, R.; Rumyantseva, A.; Lerondel, G.; Royer, P. Nano-patterning photosensitive polymers using local field enhancement at the end of apertureless SNOM tips. *J. Microsc.* **2003**, *209*, 214–222. [CrossRef] [PubMed]

6. Kirsanov, A.; Kiselev, A.; Stepanov, A.; Polushkin, N. Femtosecond laser-induced nanofabrication in the near-field of atomic force microscope tip. *J. Appl. Phys.* **2003**, *94*, 6822–6826. [CrossRef]

7. Haefliger, D.; Stemmer, A. Writing subwavelength-sized structures into aluminium films by thermo-chemical apertureless near-field optical microscopy. *Ultramicroscopy* **2004**, *100*, 457–464. [CrossRef] [PubMed]

8. Milner, A.A.; Zhang, K.; Prior, Y. Floating Tip Nanolithography. *Nano Lett.* **2008**, *8*, 2017–2022. [CrossRef] [PubMed]

9. Falcón Casas, I.; Kautek, W. In *Laser Micro-Nano-Manufacturing and 3D Microprinting*; Chapter Apertureless Scanning Near-Field Optical Lithography; Hu, A., Ed.; Springer: New York, NY, USA, 2018; in print.

10. Novotny, L.; Sánchez, E.J.; Xie, X.S. Near-field optical imaging using metal tips illuminated by higher-order Hermite–Gaussian beams. *Ultramicroscopy* **1998**, *71*, 21–29. [CrossRef]

11. Martin, Y.C.; Hamann, H.F.; Wickramasinghe, H.K. Strength of the electric field in apertureless near-field optical microscopy. *J. Appl. Phys.* **2001**, *89*, 5774–577. [CrossRef]

12. Esteban, R.; Vogelgesang, R.; Kern, K. Simulation of optical near and far fields of dielectric apertureless scanning probes. *Nanotechnology* **2006**, *17*, 475–482. [CrossRef]

13. Esteban, R.; Vogelgesang, R.; Kern, K. Tip-substrate interaction in optical near-field microscopy. *Phys. Rev. B* **2007**, *75*, 195410. [CrossRef]

14. Huang, S.M.; Hong, M.H.; Lu, Y.F.; Luk'yanchuk, B.S.; Song, W.D.; Chong, T.C. Pulsed-laser assisted nanopatterning of metallic layers combined with atomic force microscopy. *J. Appl. Phys.* **2002**, *91*, 3268–3274. [CrossRef]

15. Bouhelier, A.; Beversluis, M.; Novotny, L. Applications of field-enhanced near-field optical microscopy. *Ultramicroscopy* **2004**, *100*, 413–419. [CrossRef] [PubMed]

16. Roth, R.M.; Panoiu, N.C.; Adams, M.M.; Osgood, R.M.; Neacsu, C.C.; Raschke, M.B. Resonant-plasmon field enhancement from asymmetrically illuminated conical metallic-probe tips. *Opt. Express* **2006**, *14*, 2921–2931. [CrossRef] [PubMed]

17. Hartschuh, A. Tip-Enhanced Near-Field Optical Microscopy. *Angew. Chem. Int. Ed.* **2008**, *47*, 8178–8191. [CrossRef] [PubMed]

18. Huth, F.; Chuvilin, A.; Schnell, M.; Amenabar, I.; Krutokhvostov, R.; Lopatin, S.; Hillenbrand, R. Resonant Antenna Probes for Tip-Enhanced Infrared Near-Field Microscopy. *Nano Lett.* **2013**, *13*, 1065–1072. [CrossRef] [PubMed]

19. Wang, L.; Xu, X.G. Scattering-type scanning near-field optical microscopy with reconstruction of vertical interaction. *Appl. Phys. Lett.* **2015**, *6*, 8973. [CrossRef] [PubMed]

20. Møller, S.H.; Vester-Petersen, J.; Nazir, A.; Eriksen, E.H.; Julsgaard, B.; Madsen, S.P.; Balling, P. Near-field marking of gold nanostars by ultrashort pulsed laser irradiation: Experiment and simulations. *Appl. Phys. A* **2018**, *124*, 210. [CrossRef]

21. Gerstner, V.; Thon, A.; Pfeiffer, W. Thermal effects in pulsed laser assisted scanning tunneling microscopy. *J. Appl. Phys.* **2000**, *87*, 2574–2580. [CrossRef]

22. Milner, A.A.; Zhang, K.; Garmider, V.; Prior, Y. Heating of an Atomic Force Microscope tip by femtosecond laser pulses. *Appl. Phys. A* **2010**, *99*, 1–8. [CrossRef]

23. Huber, C.; Prior, Y.; Wolfgang, K. Laser-induced cantilever behaviour in apertureless scanning near-field optical microscopes. *Meas. Sci. Technol.* **2014**, *25*, 075604. [CrossRef]
24. Kim, K.; Song, B.; Fernández-Hurtado, V.; Lee, W.; Jeong, W.; Cui, L.; Thompson, D.; Feist, J.; Reid, M.T.H.; García-Vidal, F.J.; et al. Radiative heat transfer in the extreme near field. *Nature* **2015**, *528*, 387–391. [CrossRef] [PubMed]
25. Song, B.; Ganjeh, Y.; Sadat, S.; Thompson, D.; Fiorino, A.; Fernández Hurtado, V.; Feist, J.; García Vidal, F.J.; Cuevas, J.C.; Reddy, P.; et al. Enhancement of near-field radiative heat transfer using polar dielectric thin films. *Nat. Nanotechnol.* **2015**, *10*, 253–258. [CrossRef] [PubMed]
26. Kloppstech, K.; Könne, N.; Biehs, S.A.; Rodriguez, A.W.; Worbes, L.; Hellmann, D.; Kittel, A. Giant heat transfer in the crossover regime between conduction and radiation. *Nat. Commun.* **2017**, *8*. [CrossRef] [PubMed]
27. Dutoit, B.; Zeisel, D.; Deckert, V.; Zenobi, R. Laser-Induced Ablation through Nanometer-Sized Tip Apertures: Mechanistic Aspects. *J. Phys. Chem. B* **1997**, *101*, 6955–6959. [CrossRef]
28. Zhang, W.; Cui, X.; Martin, O.J.F. Local field enhancement of an infinite conical metal tip illuminated by a focused beam. *J. Raman Spectrosc.* **2009**, *40*, 1338–1342. [CrossRef]
29. Mihaljevic, J.; Hafner, C.; Meixner, A.J. Simulation of a metallic SNOM tip illuminated by a parabolic mirror. *Opt. Express* **2013**, *21*, 25926–25943. [CrossRef] [PubMed]
30. Huber, C.; Trügler, A.; Hohenester, U.; Prior, Y.; Kautek, W. Optical near-field excitation at commercial scanning probe microscopy tips: a theoretical and experimental investigation. *Phys. Chem. Chem. Phys.* **2014**, *16*, 2289–2296. [CrossRef] [PubMed]
31. Verhoef, A.; Zhu, L.; Israelsen, S.M.; Grüner-Nielsen, L.; Unterhuber, A.; Kautek, W.; Rottwitt, K.; Baltuška, A.; Fernández, A. Sub-100 fs pulses from an all-polarization maintaining Yb-fiber oscillator with an anomalous dispersion higher-order-mode fiber. *Opt. Express* **2015**, *23*, 26139–26145. [CrossRef] [PubMed]
32. Sader, J.E.; Borgani, R.; Gibson, C.T.; Haviland, D.B.; Higgins, M.J.; Kilpatrick, J.I.; Lu, J.; Mulvaney, P.; Shearer, C.J.; Slattery, A.D.; et al. A virtual instrument to standardise the calibration of atomic force microscope cantilevers. *Rev. Sci. Instrum.* **2016**, *87*, 093711. [CrossRef] [PubMed]
33. Hohenester, U.; Trügler, A. MNPBEM—A Matlab toolbox for the simulation of plasmonic nanoparticles. *Comput. Phys. Commun.* **2012**, *183*, 370–381. [CrossRef]
34. Available online: http://physik.uni-graz.at/mnpbem/ (accessed on 3 March 2018).
35. Green, M.A. Self-consistent optical parameters of intrinsic silicon at 300K including temperature coefficients. *Sol. Energy Mater. Sol. Cells* **2008**, *92*, 1305–1310. [CrossRef]
36. Yakubovsky, D.I.; Arsenin, A.V.; Stebunov, Y.V.; Fedyanin, D.Y.; Volkov, V.S. Optical constants and structural properties of thin gold films. *Opt. Express* **2017**, *25*, 25574–25587. [CrossRef] [PubMed]
37. Karakouz, T.; Holder, D.; Goomanovsky, M.; Vaskevich, A.; Rubinstein, I. Morphology and Refractive Index Sensitivity of Gold Island Films. *Chem. Mater.* **2009**, *21*, 5875–5885. [CrossRef]
38. Link, S.; Burda, C.; Nikoobakht, B.; El-Sayed, M.A. Laser-Induced Shape Changes of Colloidal Gold Nanorods Using Femtosecond and Nanosecond Laser Pulses. *J. Phys. Chem. B* **2000**, *104*, 6152–6163. [CrossRef]
39. González-Rubio, G.; Guerrero-Martínez, A.; Liz-Marzán, L.M. Reshaping, Fragmentation, and Assembly of Gold Nanoparticles Assisted by Pulse Lasers. *Acc. Chem. Res.* **2016**, *49*, 678–686. [CrossRef] [PubMed]
40. Lin, Z.; Leveugle, E.; Bringa, E.M.; Zhigilei, L.V. Molecular Dynamics Simulation of Laser Melting of Nanocrystalline Au. *J. Phys. Chem. C* **2010**, *114*, 5686–5699. [CrossRef]
41. Kerse, C.; Kalaycıoğlu, H.; Elahi, P.; Çetin, B.; Kesim, D.K.; Akçaalan, Ö.; Yavaş, S.; Aşık, M.D.; Öktem, B.; Hoogland, H.; et al. Ablation-cooled material removal with ultrafast bursts of pulses. *Nature* **2016**, *537*, 84–88. [CrossRef] [PubMed]
42. Finger, J.; Reininghaus, M. Effect of pulse to pulse interactions on ultra-short pulse laser drilling of steel with repetition rates up to 10 MHz. *Opt. Express* **2014**, *22*, 18790–18799. [CrossRef] [PubMed]

nanomaterials

MDPI

Article

Deformation Behavior of Foam Laser Targets Fabricated by Two-Photon Polymerization

Ying Liu [1], John H. Campbell [2], Ori Stein [3], Lijia Jiang [1], Jared Hund [3] and Yongfeng Lu [1,*]

[1] Department of Electrical and Computer Engineering, University of Nebraska-Lincoln, Lincoln,
    NE 68588-0511, USA; liuying900120@gmail.com (Y.L.); li.jia.jiang1985@gmail.com (L.J.)
[2] Material Science Solutions, 2136 Westbrook Lane, Livermore, CA 94550, USA; campbelljh@comcast.net
[3] Schafer Livermore Lab, 303 Lindbergh Avenue, Livermore, CA 94551, USA; ostein@belcan.com (O.S.);
    jhund@belcan.com (J.H.)
*   Correspondence: ylu2@unl.edu; Tel.: +402-472-8323

Received: 31 May 2018; Accepted: 3 July 2018; Published: 6 July 2018

Abstract: Two-photon polymerization (2PP), which is a three-dimensional micro/nano-scale additive manufacturing process, is used to fabricate component for small custom experimental packages ("targets") to support laser-driven, high-energy-density physics research. Of particular interest is the use of 2PP to deterministically print millimeter-scale, low-density, and low atomic number (CHO) polymer matrices ("foams"). Deformation during development and drying of the foam structures remains a challenge when using certain commercial acrylic photo-resins. Acrylic resins were chosen in order to meet the low atomic number requirement for the foam; that requirement precludes the use of low-shrinkage organic/inorganic hybrid resins. Here, we compare the use of acrylic resins IP-S and IP-Dip. Infrared and Raman spectroscopy are used to quantify the extent of the polymerization during 2PP vs. UV curing. The mechanical strength of beam and foam structures is examined, particularly the degree of deformation that occurs during the development and drying processes. The magnitude of the shrinkage is quantified, and finite element analysis is used in order to simulate the resulting deformation. Capillary drying forces during development are shown to be small and are likely below the elastic limit of the foam log-pile structures. In contrast, the substantial shrinkage in IP-Dip (~5–10%) causes large shear stresses and associated plastic deformation, particularly near constrained boundaries and locations with sharp density transitions. Use of IP-S with an improved writing procedure results in a marked reduction in deformation with a minor loss of resolution.

Keywords: two-photon polymerization; low-density foam structures; laser targets; structure deformation; acrylate resin; Raman microspectroscopy

---

1. Introduction

Two-photon polymerization (2PP) is a direct-write technology that has recently been used to create millimeter-scale laser target components to support the Department of Energy's (DOE) High Energy Density (HED) research programs [1–4]. In the first published work in this area, Bernat et al. [5] and Jiang et al. [6] report the use of 2PP to print simulated fill tubes and low-density foam-like structures, respectively. More recently, Jiang et al. [7,8], Stein et al. [9], and Oakdale et al. [10] discuss details of the design, fabrication, characterization, and assembly of low-density foam targets.

Details of the 2PP process and technology have been reviewed recently [11]. In brief, polymerization is initiated by the simultaneous absorption of two photons by a photoinitiator in a reactive monomer/oligomer resin, and it thus depends on the square of the laser irradiance. In practice, an initiator is selected that has negligible absorption at the incident fundamental laser frequency but measurable two-photon absorption at the second harmonic. Because two-photon absorption cross-sections are very low, the probability of reaction initiation is negligible except near the laser focus. Therefore, photopolymerization only occurs at the peak of the focal irradiance and generates a volumetric polymer dot ("voxel") that is generally smaller than the diffraction-limited spot size.

Typically, voxels range from 200–400 nm for fabricated structures that are similar to the ones reported here [6–8]. Structures are created by moving the repetition (rep)-rated laser beam (kHz to MHz) through the resin, thus generating overlapping voxels that, with proper scanning control, are built into the computer-aided design (CAD) three-dimensional (3D) shapes. Any unreacted resin is later removed during a post-writing development process, leaving behind a polymeric replica of the CAD design. Figure 1 shows examples of 2PP log-pile foam structures generated in our laboratory for laser target fabrication applications.

Structures described in this paper were made using one of two commercial resins: IP-Dip or IP-S. The IP-series are a family of proprietary acrylic resins plus initiator marketed by Nanoscribe GmbH for use with their commercial 2PP writing system. Using commercial resins is appealing, as it eliminates the need for custom formulation. In addition, acrylics are acceptable target materials as the atomic composition is largely carbon and hydrogen with minor amounts of oxygen. Many HED physics experiments, using foam or other polymer structures, require materials comprised of low atomic number elements because of the well-known variation in X-ray absorption (opacity) with atomic numbers [12]. Thus, the use of other common resins, such as hybrid organic/inorganic resins, for example, Ormocers, SZ2080 [13] or thiol-based resins [14], is not acceptable. Acrylic based low-density components and targets produced by 2PP are now being shot at major HED national research centers in the USA, including Lawrence Livermore National Laboratory (LLNL), Laboratory for Laser Energetics (LLE), Naval Research Lab (NRL), etc.

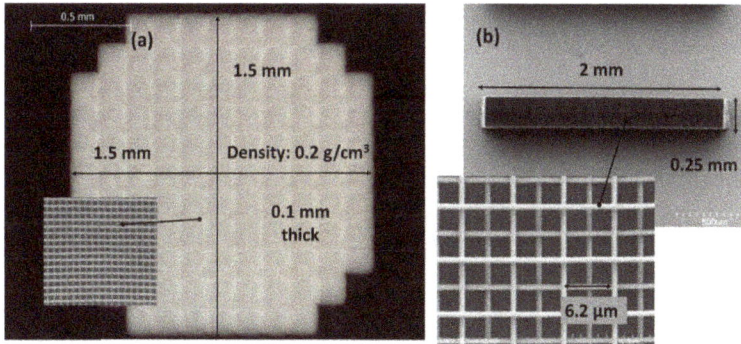

Figure 1. Examples of (**a**) foam plate and (**b**) a rod of two-photon polymerization (2PP) fabricated log-pile structures for laser target applications. The foam plate: $1.5 \times 1.5 \times 0.10$ mm^3 with a $4 \times 4 \times 2$ µm^3 beam lattice structure (density ~0.2 g/cm^3). The foam rod: $2.0 \times 0.25 \times 0.35$ mm^3 with a $6.2 \times 6.2 \times 1.0$ µm^3 beam lattice structure (density ~0.1 g/cm^3).

Acrylic resins, like IP-S and IP-Dip, are commonly used for 2PP fabrication of nano/microstructures. However, they are often limited by various materials and processing issues:

- inherent strength of the polymerized resin [14,15];
- strong sensitivity to writing conditions (peak irradiance and shots per site, i.e., dose) [16,17];
- low photo-conversion of resin to polymer [18–20];
- shrinkage during photopolymerization or development or both [6,9,20];
- stresses due to capillary forces during drying [6,9,21]; and,
- control of adhesion to the substrate [6,9].

Consequently, the user must consider each of these issues when selecting a given resin, the writing conditions, and the development method for a particular application. In this paper, we examine these issues by a combination of experiments and modeling, and suggest some possible methods for controlling the negative impacts of each in IP-Dip and IP-S resins. IP-Dip is particularly problematic. In contrast, IP-S shows significant improvement over IP-Dip, while still maintaining the required high CH content for target applications. The trade-off is in the line resolution, which is better in IP-Dip than in IP-S.

Shrinkage and deformation problems with the use of IP-Dip have been shown to stem largely from the low resin-to-polymer conversion (<50%) and an associated low modulus and yield strength of the polymer. Infrared and Raman spectroscopy are used to quantify the resin-to-polymer conversion by following the signature alkene vibrational bands. Finite element analysis (FEA) is used to simulate the degree of shrinkage and the resulting plastic strains that occur during the development and drying of 2PP log-pile foam structures.

2. Experimental

2.1. Photo-Resins and Properties

IP-Dip and IP-S commercial negative-tone, acrylate-based photoresists were used in this work (Nanoscribe GmbH, Eggenstein-Leopoldshafen, Germany), as they meet the low atomic number requirement for the proposed HED application. These resins are designed for use with Nanoscribe's Dip-in Laser Lithography (DiLL) technology and they serve as both the immersion and photosensitive material. Specifically, the resins match the refractive index of the final focusing lens and achieve the highest numerical aperture (i.e., the best resolution) at a given magnification.

IP-Dip has a low viscosity and is recommended for use in high-resolution applications requiring narrow line width. In contrast, IP-S is more viscous and designed for mesoscale printing at larger line widths. Elemental compositions and key properties of the resins are summarized in Table 1. The elemental compositions were determined using the procedure as reported in [6]. The resins are predominately CH_x with small amounts of O and traces of N thus satisfying the low atomic number requirement.

Table 1. Composition and key properties of IP-Dip and IP-S resins. Unless otherwise noted, the physical and the mechanical properties are from Nanoscribe GmbH.

Elemental Analysis of Resins					
Resin	Carbon (at.%)	Hydrogen (at.%)	Nitrogen (at.%)	Oxygen (at.%)	Empirical Formula
IP-Dip	40.2	46	0.04	13.7	$CH_2N_{0.001}O_{0.34}$
IP-S	31.5	54.1	5.8	11.8	$CH_{1.72}N_{0.086}O_{0.37}$
Physical and Mechanical Properties					
Resin	Density (liq) (g/cm^3)	Density (s)(g/cm^3) *	Young's Modulus (GPa)	Hardness (MPa)	Poisson's Ratio ***	Refractive Index
IP-Dip	1.14–1.19	1.2	0.75–2.5 **, 4.5	152	0.35	1.52
IP-S	1.16–1.19	1.2	4.6	160	0.35	1.48

* [6], ** [16], *** [22].

2.2. 2PP Microfabrication

Micro-fabrications were carried out using a Photonic Professional GT system (Nanoscribe GmbH [19]). Two-photon excitation was accomplished using the 780 nm frequency-doubled output from an Er-fiber laser (1580 nm, TEM_{00}, M^2 < 1.2, Toptica Photonics AG, Germany) operating at 80 MHz with a temporal pulse length ~100 fs. An integrated set of beam transport optics directed the laser output with circular polarization to a final focusing objective that dipped directly into the photoresist.

In the case of the IP-Dip resin, a 63X objective with a 1.4 numerical aperture was used for printing; whereas, with IP-S, the objective was 25X with a numerical aperture of 0.8. Table 2 provides a summary of the 2PP writing conditions used to fabricate the structures reported here.

Table 2. Top level summary of typical 2PP writing conditions used to prepare low-density structures reported in this work.

Parameter	Units	IP-DIP	IP-S
Final focusing power		63X	25X
Numerical aperture (NA)		1.4	0.8
Refractive index		1.52	1.48
Wavelength	μm	0.78	0.78
Beam waist (calculated)	μm	0.27	0.46
Focal spot area (calculated)	μm^2	0.23	0.66
Pulse energy	nJ	0.19	0.21
Pulse length	fs	100	100
Pulse peak power	kW	1.9	2.1
Peak irradiance	kW/μm^2	8.2	3.2
Pulse repetition rate	MHz	80	80
Average power	mW	15	17
Scan speed	μm/s	10,000	10,000
Line width (at 1cm/s scan)	μm	0.4	0.65
Shots/micron scanned		~8000	~8000

The average incident laser power was measured by a photodiode that was located at the input to the focusing objective. The passive losses in beam propagation to the sample plane were assumed to be constant and accounted for in Nanoscribe's as-built system calibration. The laser output power was controlled by an acousto-optic modulator that can be adjusted over a range of approximately 0 to 50 mW (average power). The beam diameter at focus, D_b, was calculated by:

$$D_b = \frac{2n\lambda}{\pi NA},$$ (1)

where NA is the numerical aperture, n is the refractive index and λ is the wavelength.

Structures were created using the Nanoscribe built-in software package, DESCRIBE™ 2.5 (Nanoscribe GmbH, Germany), which generates General Writing Language (GWL) files directly. The Photonic Professional GT system (Nanoscribe GmbH, Germany) uses both a piezo stage and two coupled galvanic mirrors to write the structure. The galvanic mirrors allow for rapid x-y scanning at up to 10–20 mm/s over an area 200 μm in diameter when using the 63X final focusing objective or 400 μm in diameter when using the 25X objective lens. The vertical (z) motion is controlled by the piezo stage and the built-in z-drive of the focusing objective, which ranges up to several mm in height. The system can print structures with an area of up to 25 × 25 mm^2 by using the motorized stage and "stitching" the structures together. The largest dimension of the structures that were fabricated in this application was 2 mm. The stitching accuracy is typically 1–4 microns [6].

The photoresist was deposited as a drop on a 25 × 25 × 0.7 mm^3 glass substrate that was mounted in an aluminum sample tray. The tray, with substrate and resist, was inserted into the Nanoscribe GT housing that is attached to a precision piezoelectric-driven stage. All of the operations were carried out under yellow room lighting to avoid polymerization by single-photon absorption.

2.3. Structure Development, Drying, and Characterization

After exposure, the sample substrate with IP resin was removed from the holder and was developed at room temperature for 1 h in 50 mL of propylene glycol monomethyl ether acetate (PGMEA, Sigma-Aldrich, St. Louis, MO, USA), followed by a 1 h soak in 25 mL of isopropyl alcohol (IPA, Sigma-Aldrich, St. Louis, MO, USA). If a release layer was used, then the substrate plus the structure were removed from the IPA and immersed in the Microchem-specified release agent (Remover PG™, MicroChem Corp., Newton, MA, USA). The IPA was removed for either air or supercritical drying. Supercritical drying was accomplished using carbon dioxide (CO_2) and a commercial drying system (SPI-DRY™, SPI Supplies, Inc., West Chester, PA, USA). Completion of the IPA solvent exchange with CO_2 was determined using gas chromatography. The exchange was terminated when the residual IPA attained a level of 0.03% in the monitored CO_2 effluent.

The surface morphology was characterized by optical and scanning electron microscopy (SEM). SEM images were obtained using a Hitachi model S4700 (Hitachi, Ltd., Tokyo, Japan). To obtain high-quality images, the samples were vapor coated with ~5 nm of chromium or gold. The imaging voltage was kept low (<10 kV) to avoid damaging the structures.

Raman spectra were recorded using a Raman microscope (Renishaw, InVia™ H 18415, UK) operating at an excitation wavelength of 785 nm and was focused onto the sample through a 50X objective lens (NA 0.75). Raman scattering was collected using the same lens. The average laser power and accumulation time used to record the Raman spectra were 10 mW and 10 s, respectively. Fourier transform infrared (FTIR) spectra were recorded on resins and polymerized thin films between 400 and 4000 cm^{-1} using a FTIR spectrometer (Nicolet™ iS50, Thermo Fisher Scientific Co., Waltham, MA, USA) equipped with diamond attenuated total reflection (ATR). Polymerized thin films (6 μm) were prepared by spin coating the resin on fused silica substrates, and then curing by single-photon polymerization at 395 nm for 10 min at 12 mW/cm^2.

2.4. Finite Element Analysis

Finite element analysis was used to simulate the shrinkage and deformation of the log-pile structures and foam rods via COMSOL Multiphysics® 5.3 software (COMSOL, Inc., Burlington, MA, USA), assuming a linear elastic response. Mesh configurations were created using COMSOL's built in "fine mesh" to ensure solution convergence and computational efficiency. Constrained (fixed) boundary conditions were chosen to simulate the adhesion of the polymerized resin to the substrate.

Input material properties are given in Table 1. The effective Young's modulus for the open cell foam, E_f, was estimated using the correlation, as reported by Ashby [23]:

$$E_f = E_s \left(\frac{\rho_f}{\rho_s}\right)^2, \tag{2}$$

where, ρ_f is the density of foam; and, ρ_s and E_s are the density and Young's modulus, respectively, for the polymerized resin.

Shrinkage was simulated by an equivalent thermal contraction of the structure using a stepped temperature drop and a user-defined thermal expansion coefficient and heat capacity for the foam and top layer. In contrast, a standard solid mechanics treatment was used to model the deformation caused by capillary forces. Further details are given in Section 3.5.

3. Results and Discussion

3.1. Structural Resolution of ID-Dip and IP-S Resin

The feature size of a microstructure fabricated by 2PP is determined by the size of the voxels, which is related to the induced photon intensity and sequent chemical reactions. The absorption of photons depends on the square of the light intensity, and the use of ultrashort pulses can start intense nonlinear processes at relatively low average power [24]. Theoretical studies have been established by several groups to investigate the dependence of linewidth that is based on nonlinear absorption [25,26].

In experiment, measurements of 2PP line widths versus laser power have been reported for three Nanoscribe resins (IP-DIP, -L780, and -G780) [6]. The work was carried out using the same Nanoscribe Professional GT system used here and at scan rates of 10 and 20 mm/s. A simple engineering model was used to predict the line characteristics vs. laser power and scan rate. Here, similar measurements and treatment are reported for IP-S.

A set of support bars was first printed followed by a series of suspended lines normal to the bars (Figure 2a). The lines were printed using different laser powers and suspended to avoid complications due to interactions at the resin-to-glass interface. Each laser pulse above the threshold power initiated some degree of polymerization in an ellipsoidal-shaped voxel at laser focus. The fast laser repetition rate generated a continuous line of polymer comprised of closely overlapping voxels, each having an effective volume, V_{vox}. The typical spacing between successive shots was ~0.1 nm at a scan rate of 10 mm/s and a laser repetition rate of 80 MHz, so the volume within a typical effective voxel received ~10^3–10^4 laser shots (Table 2).

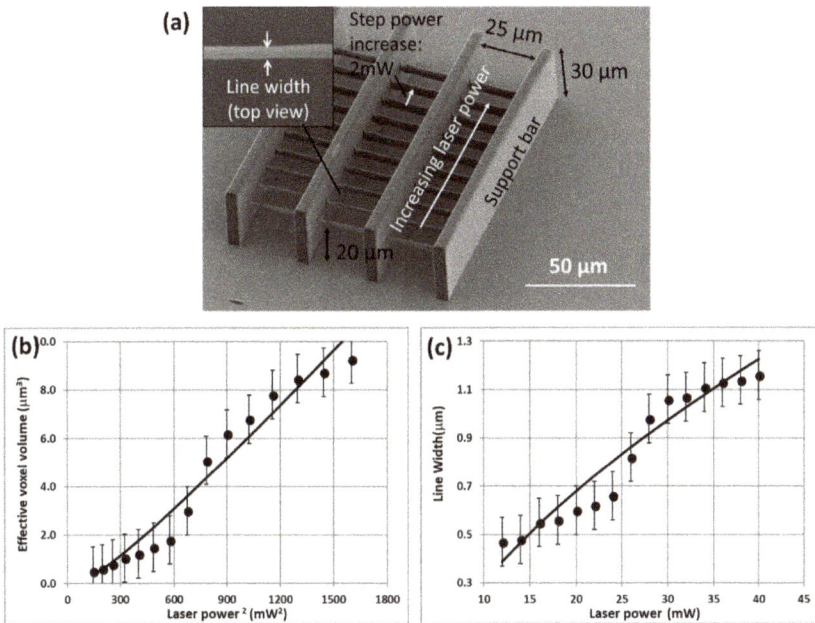

Figure 2. (**a**) Suspended line structures used to quantify 2PP line width vs. laser power for IP-S. Line widths were measured by scanning electron microscopy (SEM) (at normal incidence; see inset image) for lines printed in 2.0 mW stepped-increments of laser power; (**b**) Measured and calculated effective voxel volume vs. laser power^2 and (**c**) linewidth vs. laser power for IP-S resin. The lines were calculated using the model described in Equations (3)–(6).

The results are plotted in Figure 2b,c in terms of the effective voxel volume and linewidth as a function of the average laser power. The data were analyzed using a simplified engineering treatment that was initially suggested by Leatherdale and DeVoe [27], and more recently used by Thiel et al. [28]. These authors relate the absorbed dose in an initiated voxel volume, V_i (nm^3), to the laser operating conditions:

$$V_i \sim k\,(P_a - P_t)^2\,t_{exp}, \tag{3}$$

where P_a is the laser average power (mW), P_t is the threshold power (mW), t_{exp} is the exposure time (s), and k is a proportional constant. The square dependence on power is due to the two-photon nature of the process and, thus directly proportional to the square of the per-pulse peak laser irradiance above the threshold. Note that we report the results in terms of the system average output power, rather than peak irradiance to simplify comparison with the typical system operational parameters. The average laser power was monitored and controlled during system operation.

Making use of the fact that the exposure time was inversely proportional to the scan rate (R_s) and assuming the effective polymerized voxel volume, V_{vox} (nm³), was proportional to the initiated volume, V_i, led to:

$$V_{vox} \sim k'(P_a - P_t)^2 / R_s, \tag{4}$$

Experiments showed the polymerized voxel was ellipsoidal with diameter, D, and length, Z, perpendicular and parallel to the beam propagation direction, respectively, giving a geometric volume:

$$V_{vox} = \frac{\pi D^2 Z}{3} = \frac{\pi D^3 A_r}{3}, \tag{5}$$

where A_r is the line aspect ratio, Z/D, which for IP-Dip and IP-S was measured at 2.5 and 5.4, respectively. Note that the ratio of the aspect ratios for IP-S/IP-Dip is 2.2, in good agreement with the value of 1.8 for the ratio of the numerical apertures that were used for printing in IP-S (NA = 1.4) and IP-Dip (NA = 0.8).

Combining Equations (4) and (5) and recognizing that the linewidth (L_w) equals the effective voxel diameter (D) under constant scan rate conditions leads to the useful correlation for linewidth vs. operating laser power:

$$L_w = \left[k' \frac{3(P_a - P_t)^2}{\pi A_r R_s} \right]^{\frac{1}{3}}, \tag{6}$$

The measured and calculated effective voxel volume and linewidth for IP-S are plotted vs. $(P_a - P_t)^2$ and $(P_a - P_t)$ in Figure 2b,c, respectively. In general, the agreement is reasonable given the indicated error in linewidth measurements. Similar reasonably good agreement has been reported in prior tests using IP-Dip and other resins [6].

A threshold power (P_t) of 6 mW was assumed for IP-S, which is equivalent to the value for IP-Dip and other resins that were determined at very low scan rates (~0.1 mm/s) [20]. This threshold agreed with the lack of detectable polymerization (i.e., lines) below 12.5 mW at the much greater scan rates used here (10 mm/s). Clearly, some degree of polymerization (gelation) occurs at powers between 6–12 mW; but the dose is insufficient to generate a structure that is capable of surviving development.

In certain cases, the polymerization rate can be varied by a change of laser power and scan speed. Some authors have reported changes in polymerization propagation and termination rates due to temperature gradients formed around the focal point during 2PP [29,30]. However, in situ temperature measurements have not revealed a significant heating effect on the polymerization process when working at close-to-threshold conditions [31]. Therefore, the effect of localized thermal accumulation on 2PP fabricated structures is not included in this fitting model.

The primary benefit of the analysis reported here is as an engineering tool that, by interpolation, can reliably predict the line dimensions at different laser operating conditions in a given resin. The major limitation is that the analysis is largely an empirical treatment and it does not address the details of the excitation and complex polymerization chemistry of the process.

3.2. Plastic Strain in Simple Beam Structures Written in IP-Dip and IP-S Resins

Polymers characteristically have low elastic moduli and yield strengths but can accommodate significant plastic strain before ultimate failure [32]. These characteristics are an advantage in many 2PP applications. For example, many photo-resins undergo some shrinkage during polymer conversion, as evidenced by the greater density of the polymer vs. resin phase. Typically, polymer shrinkage is less than 2%. Consequently, polymers can generally accommodate small amounts of plastic strain without failure, thus leaving the desired structure fully intact.

Problems tend to arise in 2PP fabrication when there are large strains, particularly in the cases of significant shrinkage or differential shrinkage during development. Examples of this are shown in Figure 3 for two log-pile like foam blocks fabricated in IP-Dip resin. The blocks were designed to be $50 \times 50 \times 50$ µm^3; yet, after drying, both had shrunk by ~10% to ~45 µm in width. The extent of the shrinkage was clearly visible in the rows at the base of the block where the fabricated lines contacted and adhered to the substrate. This degree of shrinkage is consistent with the shrinkage measured in other foam-like log-pile structures that were fabricated here and reported elsewhere [6,9,11,14].

Both structures in Figure 3 accommodated the shrinkage without evidence of plastic strain in the central portion of the structure. In such cases, one could attempt to compensate for shrinkage by simply designing and fabricating a proportionally larger structure.

Problems due to shrinkage were most evident at the boundaries of IP-Dip log-pile structures (Figure 3b). Large shear stresses developed at the fixed boundary between the polymer and the substrate. Also, certain structural elements, such as cantilever-type beams or simple beams that span long unsupported distances, exhibited large plastic deformation. Such deformations are visible in the structure in Figure 3b, but are noticeably absent for the shorter spans in Figure 3a.

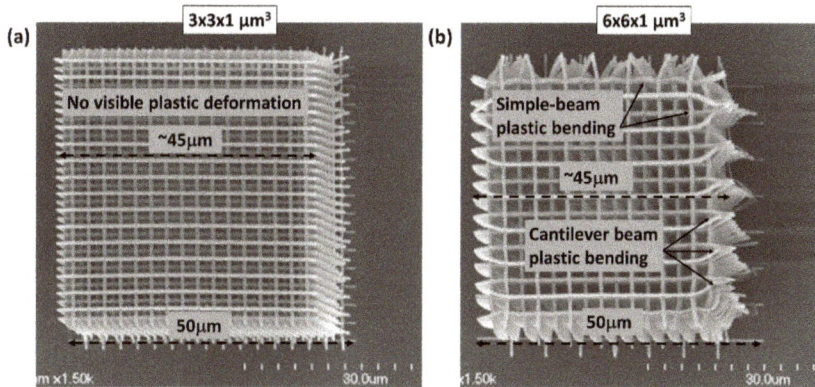

Figure 3. Log-pile structures with (**a**) $3 \times 3 \times 1$ µm^3 and (**b**) $6 \times 6 \times 1$ µm^3 cell size fabricated in IP-Dip resin with ~300–400 nm line width. Note the lack of observable plastic deformation at the smaller cell size in (**a**) in contrast to the visible bending in the simply-supported and cantilever beam sub-elements at the larger cell size in (**b**).

Similar log-pile structures were written in IP-S (Figure 4). The linewidth and height were greater than those in IP-Dip because of the larger numerical aperture, as discussed in Section 3.1. Consequently, IP-S structures had to be fabricated using a bigger cell size ($6 \times 6 \times 3$ µm^3, Figure 4a) to achieve foam densities that are equivalent to IP-Dip. The beams were laterally offset in successive layers with repeating alignment on every fourth vertical layer (Figure 4a). Each beam in the log-pile was fabricated using vertically offset and partially overlapping (50%) double scans to achieve the 3 µm height (Figure 4a inset). In general, the IP-S log-pile structures exhibited significantly less deformation than the similar structures that were written in IP-Dip (Figure 4b–d).

To better compare the strengths of structures written in IP-Dip vs. IP-S, we fabricated, developed, and air dried a series of simply supported and cantilever beams in the two resins (Figures 5 and 6). Except for the beam length, all of the structures were designed and fabricated in the same way. The beam cross-sections were designed to be 3×3 µm^2 and were fabricated using a 10-wide $\times$ 6-high scan grid, specifically 10 lateral scans at 0.3 µm line spacing and six vertical scans at 0.5 µm layer spacing. The average laser power was 15 mW, and the scanning speed was 10 mm/s. The beam structures were developed, rinsed, and air dried, as described in Section 2.3.

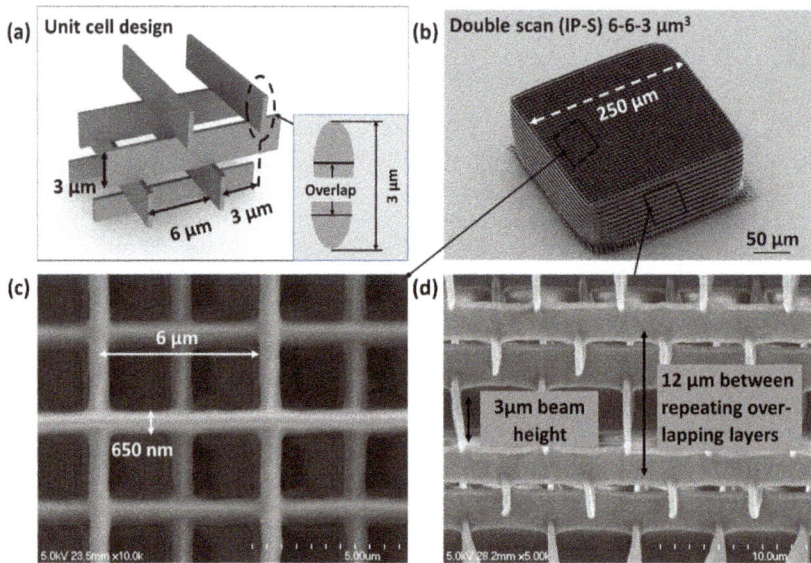

Figure 4. (**a**) The design of a log-pile structure with a 6 × 6 × 3 µm³ cell size fabricated using 50% overlapping double scans as described in the text. SEM images of (**b**) the 250 × 250 × 100 µm³ foam block fabricated in IP-S resin and in magnified views from (**c**) the top showing the linewidth and horizontal lattice spacing and (**d**) the side indicating the repeating overlap of every fourth layer, i.e., 4 × 3 um = 12 um.

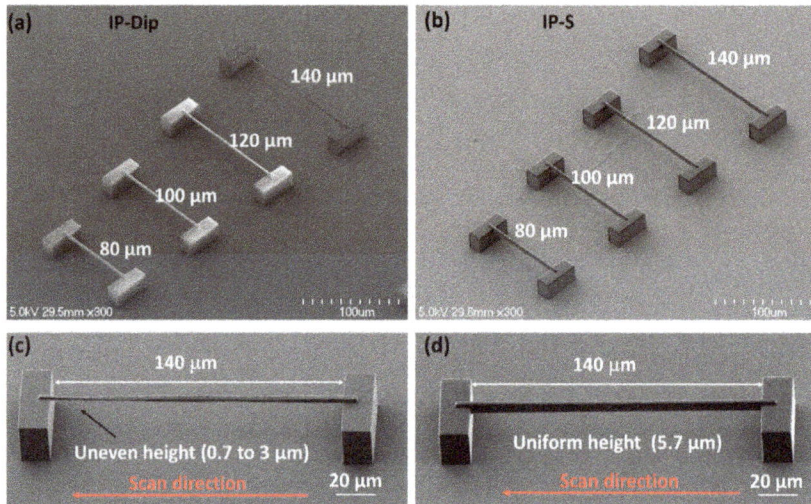

Figure 5. Simply supported beam structures fabricated in (**a,c**) IP-Dip and (**b,d**) IP-S resin. The scan direction was from right to left, as indicated by the arrow. The average laser power was 15 mW, and the scanning speed was 10 mm/s; see the text for further details.

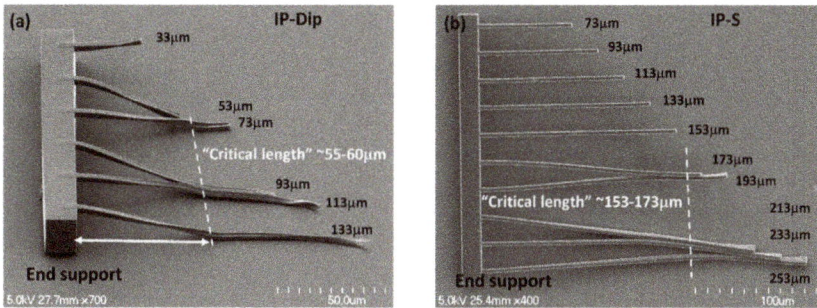

Figure 6. Cantilever beam structures of varying lengths with an integrated end support printed in (a) IP-Dip and (b) IP-S resin. The printed beam width is 3 μm with a lateral spacing between beams of ~20 μm and vertically suspended above the base substrate by ~20 μm. The "critical length" for collapse under capillary drying forces is indicated by the dashed line.

The final objectives used for IP-S and IP-Dip were 25X (NA = 0.8) and 63X (NA = 1.4), respectively (Table 2), with associated fabricated linewidths of ~0.4 and ~0.65 μm. Thus, adjacent scan lines overlapped more in IP-S than in IP-Dip.

Simply supported beam structures that were fabricated in both IP-S and Dip showed no measurable plastic deformation after development and air drying (Figure 5). The only difference in performance between the two resins was (a) unevenness in the vertical thickness of the longest beam fabricated in IP-Dip (140 μm) and (b) greater overall vertical beam thickness as achieved in IP-S. The latter effect was due to the greater depth of field (Rayleigh range).

Figure 6 shows cantilever beams that were fabricated in IP-S and IP-Dip and then developed, rinsed in IPA, and air dried. Some of the longer beams plastically deformed to such an extent that they were connected in pairs as well as to the substrate. This is not surprising as the liquid meniscus that drives the capillary forces would be expected to span the spaces between the beams, as well as connect the beams to the substrate.

The effect of capillary drying forces on deformation in microscale cantilever and simply supported beams has been rather extensively studied because of the common use of air drying for solvent removal in many microfabrication processes (for example, [33–36]). The extent of deformation is generally characterized by a "critical length", which refers to the distance from the beam attachment at the end support to the point of beam adhesion to a neighboring beam, the substrate, or both. For example, the critical length that was observed for the cantilever beams that were fabricated in IP-Dip was ~55–60 μm, whereas for IP-S, the value was ~153–173 μm (Figure 6).

Liu et al. [33] and Mastrangelo and Hsu [35] both provide closed-form solutions for estimating the critical length based on the polymer properties, beam dimensions, and inter-beam spacing. Although their mathematical approaches differ, they arrive at the same relationship:

$$L_c = \left[\frac{3Ew^3d^2}{8\gamma \cos(\theta)} \right]^{1/4} \tag{7}$$

where L_c is the critical length (μm), E Young's modulus (GPa), w beam width (μm), d beam spacing (μm), γ solvent surface tension (22 dyne/cm, IPA), and θ the wetting angle. Here, we assumed the structure was fully wetted (θ ~0°). Using the reported Young's modulus for IP-S of 4.6 GPa (Table 3) gives a critical length of 157 um, which agrees well with the measured value. Repeating the same calculation for IP-Dip is problematic as Young's modulus depends strongly on the writing speed and laser power (i.e., energy dose, J/cm³). For example, Lemma et al. [16] report a linear increase of 0.35 GPa/mW in Young's modulus from ~0.75 to 3.6 GPa over a range in average laser power from 5–13 mW. The writing speed was 100 µm/s. In the work reported here, the writing speed was 10,000 um/s at a laser power of 15 mW. Therefore, Young's modulus was expected to be lower. Equation (7) was used to estimate a Young's modulus of ~0.1 GPa based on the observed cantilever beam critical length of ~60 µm (Figure 6a).

Table 3. Summary of FTIR peak intensities (normalized to the C=O peak) for CH_2=CH- stretching and bending vibrational modes.

Band (cm^{-1})	Group and Mode	IP-S: Peak Intensity			IP-Dip: Peak Intensity		
		Resin	UV-Cured Film	DC	Resin	UV-Cured Film	DC
~1635	C=C stretch	0.06	0	100	0.07	0.02	71.43
~1405	C=C bend	0.03	0	100	0.34	0.08	76.47
~940	C=C bend	0.11	0	100	N.D.	N.D.	N.D.
~810	C=C bend	0.1	0	100	0.41	0.07	82.93

Mastrangelo [35] also treats the case of capillary collapse for a simply supported beam (i.e., a beam clamped at both ends). Using his results, we predicted the critical span to be ~160 and 400 µm for IP-Dip and IP-S, respectively. This agreed with the lack of collapse that was observed for the beam structures in Figure 5.

Polarization of the laser beam has been reported to affect the intensity distribution and thermal gradients around the focal spot thus leading to different polymerization rates, which can, in certain cases, affect the feature size and introduce small changes (~20%) in some mechanical properties [30]. We believe that impact of polarization effect on mechanical properties is likely to be small for our application compared to other effects. For example, Young's modulus was estimated to be ~0.1 GPa for the IP-Dip polymerized structures written here, while the fully polymerized IP-Dip photoresist has a Young's modulus of 4.5 GPa (Table 1). Also, at the employed high writing speed (10,000 µm/s), anisotropy in heat flow would be a second order effect for enhancing the mechanical stability of the foam targets. Besides, the beam is circular polarized for our writing process. Other research work has shown that the circular polarization of incident light could ensure a more spherical voxel within the xy-plane [37]. This avoids polarization-dependent linewidth between separate log-pile layers where the scan directions are perpendicular to each other.

Yoshimoto et al. [34] offers a different approach for describing plastic yield in micro-cantilever beams, with the resulting expression for the critical length in terms of the yield strength:

$$L_c = w \left[\frac{\sigma_y d}{6\gamma \cos(\theta)} \right]^{1/2} \tag{8}$$

where δ_y is the yield strength (MPa) and the other variables are the same as given above. To our knowledge, the yield strength for IP-Dip and IP-S has not been reported, so Equation 7 and the results in Figure 6 provide a means to estimate these values, specifically, δ_y ~3 MPa for IP-Dip and ~20 MPa for IP-S.

3.3. Fourier Transform Infrared and Micro-Raman Vibrational Spectroscopy of Resin Conversion

FTIR and micro-Raman vibrational spectroscopy were used to monitor the degree of polymerization in the IP-S and IP-Dip acrylic resins. Other recent studies have shown these techniques provide a wealth of molecular detail at the nano to microscale, about the extent of monomer/oligomer photo-conversion (see, for example, [11,14,20,38,39]). The characteristic vibrational bands that were associated with the CH_2=CH-, C=O, and C-O groups that comprise the two resins are well known [40–43] and are clearly detected in both the FTIR and micro-Raman spectra (Figure 7).

FTIR bands are due to linear optical absorption by an oscillating dipole associated with the vibrations of a particular molecule or functional group [41,42]. In contrast, Raman bands are scattering phenomena and relate to the polarizability of the molecule or the molecular group. Specifically, Raman bands are associated with the radiation from an oscillating dipole that was induced by the incident laser electric field [40,41]. Thus, the two methods are complementary in that vibrational bands that are weak or not detected by one method may be detected by the other; this is often the case for our application.

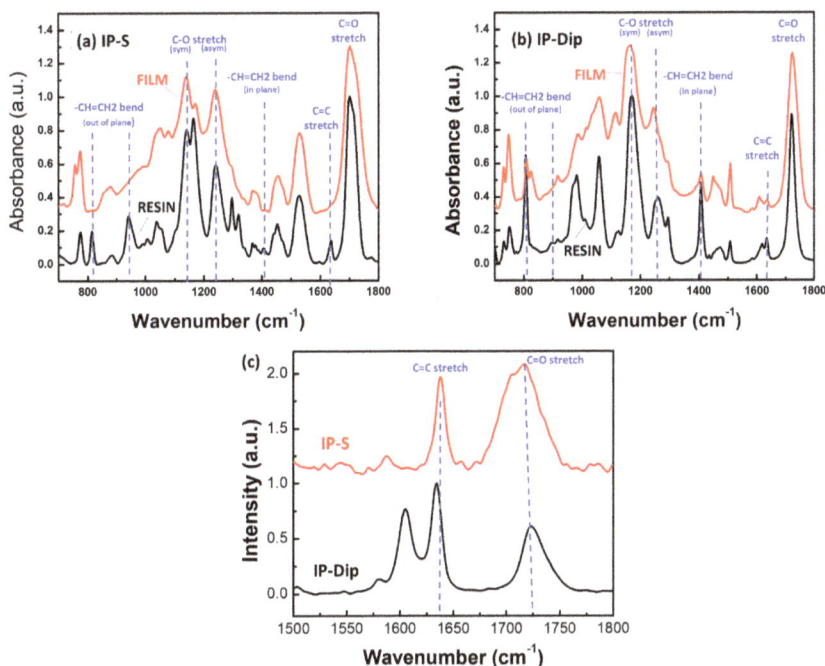

Figure 7. Fourier transform infrared (FTIR) spectra of the resin and fully cured film of (a) IP-S and (b) IP-Dip over the fingerprint region of 700–1800 cm^{-1}. The bands associated with the terminal CH$_2$=CH- stretching and bending modes are indicated on the spectra. (c) Raman spectra of IP-Dip and IP-S 2PP cured photoresists.

In general, the FTIR spectra provide greater structural detail across the so-called molecular fingerprint region (~700 to 1800 cm^{-1}), which includes characteristic stretching and bending modes of CH$_2$=CH-, C=O, and C-O [42,43]. In addition, the method is insensitive to fluorescence from the initiator in the resin. The major drawback is that the FTIR spectrometer can only probe macroscopic samples. Raman microspectroscopy, on the other hand, has the advantage of being able to probe small volumes (<0.5 μm dia.) and it has greater sensitivity to the CH$_2$=CH- stretching vibration at ~1600–1640 cm^{-1} [44]. This bond has a low dipole moment and it gives only weak FTIR bands, while the Raman signal is strong due to the large polarizability of the C=C bond. The main drawback to micro-Raman for our application was the interference caused by fluorescence from the initiator in the resin.

FTIR spectra of unreacted resins and UV-cured films of IP-S and IP-Dip are shown in Figure 7. Table 3 summarizes the measured strength of the C=C stretching band and the three bending modes at ~1635, 1405, ~810, and ~940 cm^{-1}, respectively. The intensity was normalized to the C=O band intensity because that group concentration is expected to remain constant in a given sample. The degree of conversion (DC) was calculated by comparing C=C stretching or bending mode intestines to a reference C=O band before and after UV photopolymerization [15,20]:

$$DC = \left[1 - (A_{C=C}/A_{C=O})/(A'_{C=C}/A'_{C=O})\right] \times 100, \tag{9}$$

where $A_{C=C}$, $A_{C=O}$, $A'_{C=C}$, and $A'_{C=O}$ are the integrated intensity of corresponding peaks in the polymerized and the unpolymerized resins. The IP-S spectra showed the expected result of complete reaction of the terminal alkene group after UV exposure. In contrast, the UV-exposed IP-Dip sample still contained ~17–29% unreacted C=C based on the ratios of the bands at ~810, ~1405, and 1635 cm^{-1} before and after UV exposure.

We next carried out micro-Raman measurements of lines that were written by 2PP in both resins using the conditions that are summarized in Table 2. The samples were developed and then examined using Raman microscopy, as described in Section 2.3.

Figure 7c shows the Raman bands for the C=O and CH$_2$=CH- stretching modes after 2PP exposure. Both of the resins showed significant amounts of unreacted CH$_2$=CH-. IP-Dip in particular showed low conversion, which is consistent with the trend that was observed in the UV-exposure results in Figure 7b. The reason for the difference in vinyl conversion of IP-Dip vs IP-S resins is difficult to assess without detailed knowledge of the chemical structure of these proprietary resins. Nevertheless, the IR and Raman data coupled with the elemental resin composition do offer a few hints. We assume here that the vinyl conversion continues (propagates) until the well-known free radical termination by oxygen (O$_2$) [45]. Note that, based on the IR spectra and elemental composition results, IP-Dip is a fully vinyl-acrylate resin, whereas IP-S does show the presence of some amine functionality (amine bands at ~3300–3500 cm^{-1}). We suspect that IP-Dip conversion becomes sterically hindered early on. In other words, molecular rearrangement is too slow (diffusion limited) to permit access to the unreacted vinyl groups before the small, highly mobile O$_2$ terminates the reaction. In contrast, the full UV conversion of IP-S suggests adequate mobility to achieve full reaction before termination. The incomplete conversion of IP-S under 2PP is possibly due to insufficient initiation. Of course, use of other common organic chemistry structural analysis tools, for example, C^{13} and H^1 nuclear magnetic resonance (NMR), gas chromatography-mass spectrometry (GC-MS), size exclusion chromatography (SEC), etc., could fully elucidate the structures of both resins. We have chosen not to do that here. Hopefully, the resin vendor will soon publish the structure making such analyses unnecessary.

The results suggest that low conversion is a major factor contributing to the large shrinkage and low modulus and yield strength of the IP-Dip foam structures. This agrees with prior work by Jiang et al. [12,15], who also reported low conversion by 2PP in custom resins with a resulting loss in mechanical integrity.

It is also probable that the C=C conversion varies through the width of the line, being greatest at the point of peak irradiance at the center of the focal spot and lower near the gaussian beam edges. Thus, the effective thickness of the line would be less than the observed thickness. Because the mechanical strength varies as approximately the cube of the line thickness, and then small negative changes in the effective line thickness would significantly weaken the part.

3.4. Fabrication of Foam Rods in IP-Dip and IP-S Resins

Figures 8 and 9 show foam rods (2.0 × 0.25 × 0.35 mm^3) fabricated in IP-Dip and IP-S, respectively, using the writing conditions in Table 2. Both foam structures have a design density of 0.10 g/cm^3. One of the IP-Dip rods was fabricated as a 100% foam log-pile structure (Figure 8a), whereas the other had a 15-μm-thick cap layer that functioned as a laser ablator (Figure 8b). Only IP-S rods with the 15 μm cap layer were fabricated (Figure 9).

The IP-Dip and IP-S rods were comprised of a series of 125 × 125 × 100 μm^3 and 250 × 250 × 100 μm^3 sub-blocks, respectively. The ability to print larger sub-blocks for the IP-S rods is a direct result of the numerical aperture for printing (IP-S 0.8 vs. 1.4 for IP-Dip). The IPS sub-block width was designed to match the rod width, thereby reducing the number of stitching boundaries by eightfold over that for rods that were written in IP-Dip. The foam cell size for the IP-S rods was also designed to be larger ($6 \times 6 \times 3 \ \mu m^3$) to offset the mass of the thicker beams and still achieve the 0.10 g/cm^3 density goal. Each IP-S beam was fabricated with vertically offset and partially overlapping double scans (50%) to achieve a taller and stiffer (more reacted) structure (Figure 4a).

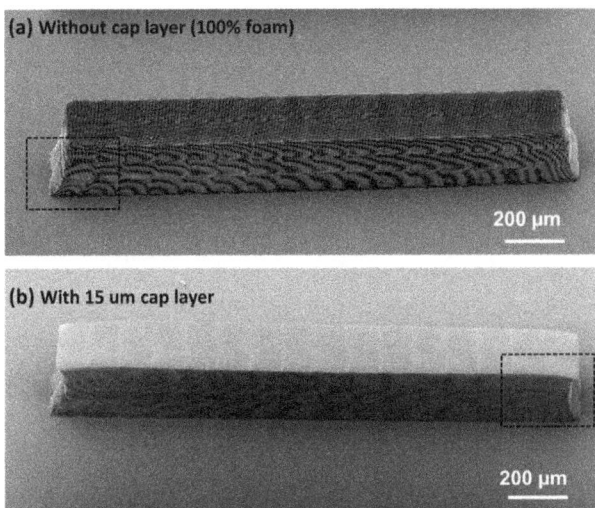

Figure 8. SEM images of $2 \times 0.25 \times 0.3 \ mm^3$ foam rod with x, y, z cell dimensions of $6.2 \times 6.2 \times 1 \ \mu m^3$ fabricated in IP-Dip (**a**) without and (**b**) with a 15-um-thick fully dense cap layer.

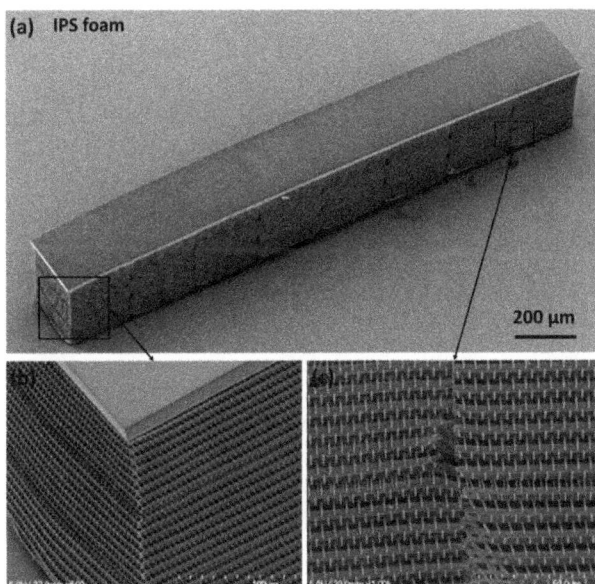

Figure 9. SEM images of (**a**) $2 \times 0.25 \times 0.3 \ mm^3$ foam rod with x, y, z cell dimensions of $6 \times 6 \times 3 \ \mu m^3$ and showing areas at higher magnification to illustrate (**b**) structure and (**c**) stitching boundary quality.

The IP–Dip rods showed the maximum deformation of ~6–7% at the interfaces of the foam with the substrate and with the solid cap layer (Figure 8). In contrast, the IP-S foam rods showed much less shrinkage and deformation in these interfaces (Figure 9). The largest defects/deformations in the IP-S rod occurred at the corners of the stitching boundaries (Figure 9b,c) and were consistent in size and location throughout the structure. Because these defects occurred at the outer surface of the rod, they did not impact the region of the target that was irradiated by the laser. Nevertheless, work is continuing on improving the process to eliminate these defects.

3.5. Analysis of Shrinkage and Deformation in IP-Dip Foam Rods

Greater detail on the deformation of the IP-Dip rods at the interfaces between the foam-substrate boundary and the full-density cap layer are shown in Figure 10. Specifically, the images in Figure 10a,c provide magnified views of the rod end regions within the dashed boxes in Figure 8.

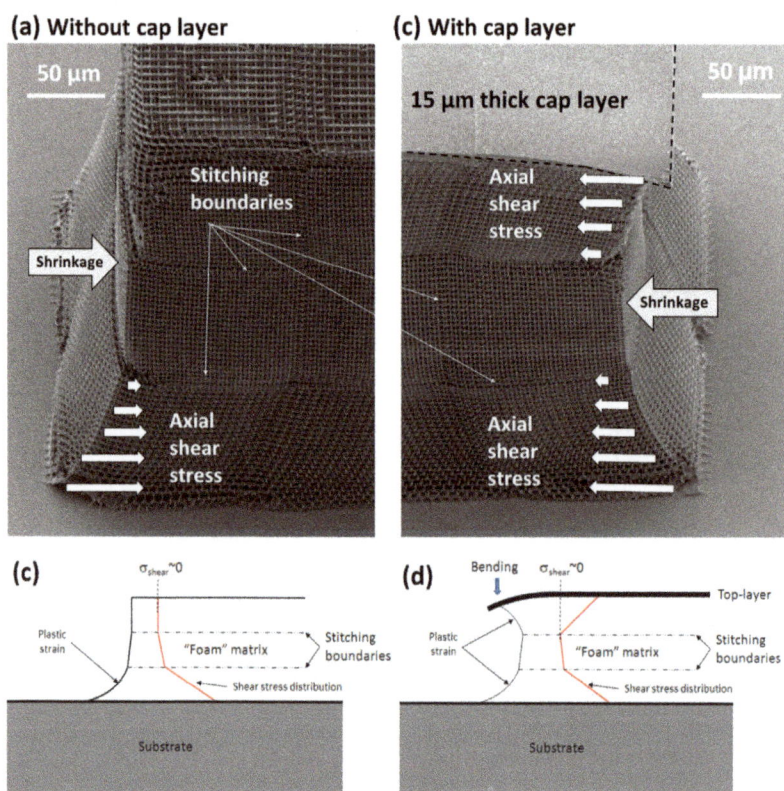

Figure 10. Details of the structural deformation of IP-Dip foam rods, specifically showing the regions within the dashed boxes in Figure 8. The SEM images show the ends of the rod (**a**) without and (**b**) with a 15-um-thick fully dense cap layer. Resin shrinkage during development and drying produced residual axial shear stresses and associated plastic strains in both the foam only and the foam with top cap, as shown schematically in (**c,d**) and also indicated in the SEM images; the arrow lengths are notional representations of the relative magnitude of the axial shear stresses.

Comparison of the designed and measured dimensions of the IP-Dip foam rods indicated ~10% maximum axial shrinkage at the top of the foam rod. This is consistent with the measured shrinkage in other foam-like structures that were fabricated here and reported elsewhere [6,9]. The images in Figure 10 show that the axial shear stresses (and strains) due to shrinkage were greatest near "fixed" boundaries that were constrained by adhesion of foam to the glass substrate and to the full density top cap. Also, the stitching boundaries (Figure 10a,c) are regions of inherently lower strength that in some cases have been observed to cause layer-to-layer delamination under shrinkage-induced shear strains.

The axial shear stresses are indicated schematically in the SEM images in Figure 10a,b where the arrow lengths are notional representations of the relative magnitude of the stresses. Shear stresses in the lateral direction also exist, but are smaller.

Deformation of IP-Dip foam rods during the development process may be due to capillary forces and/or internal shrinkage due to the complete or partial removal of unconverted resin within the fabricated lines. Simulations were used to separately investigate the magnitude of these two effects by FEA. The geometry and the mesh configuration of a unit lattice and the foam rods with and without the cap layer are shown in Figures 11a and 12a,b, respectively. The base of the lattice and the foam were constrained to simulate adhesion to an infinitely stiff substrate. In the case of the rod, the top cap was also assumed to fully adhere to the foam, although each can elastically respond to the applied stress.

We first considered the effects of meniscus drying forces. The capillary pressure, P, due to meniscus forces during air drying, can be estimated using the Young–Laplace equation:

$$P \sim 4\gamma \cos(\theta)/L \tag{10}$$

where L is the effective pore diameter (or inter-beam distance), γ is the solvent surface tension, and θ is the wetting angle. Applying Equation (10) to the $6.2 \times 6.2 \times 1$ µm³ foam unit cell and using a surface tension for IPA of $\gamma = 22$ dyne/cm and assuming a fully wetted structure ($\theta \sim 0°$) gives an estimated maximum capillary pressure of ~8 kPa (~12 psi). The FEA of the foam rods predicted a maximum three-dimensional (3D) deformation of only ~130 nm over the entire length of the rod structure. This was less than half of the 2PP line width. Consequently, the deformation was expected to be well within the elastic limit with no permanent plastic strain, even at the very low estimated foam modulus value of ~0.1 MPa.

In contrast to the capillary drying effects, FEA simulations of the impact of shrinkage showed deformations that closely matched observations. Figure 11 shows show the FEA mesh configuration and the simulations results for a $24 \times 24 \times 8$ µm³ beam lattice representing the microscale details of an individual building block of the foam rod. The lattice architecture had a horizontal beam (x, y) spacing of 6.2 µm, a beam height (z) of 1 µm, and a beam width (i.e., line width) of 0.4 µm. The simulation showed that shrinkage of the log-pile structures leads to uniform compression in the center region and plastic deformation at the end, as is consistent with the behavior shown in Figure 3. The vertical strain gradient (Figure 11b,c) was ~6–7 nm/micron over the 8 µm height and 24×24 µm base of this lattice structure. This predicted an expected total deformation of ~35 µm over the ~60 µm thick first layer of the 2 mm foam rod (Figure 10a,b).

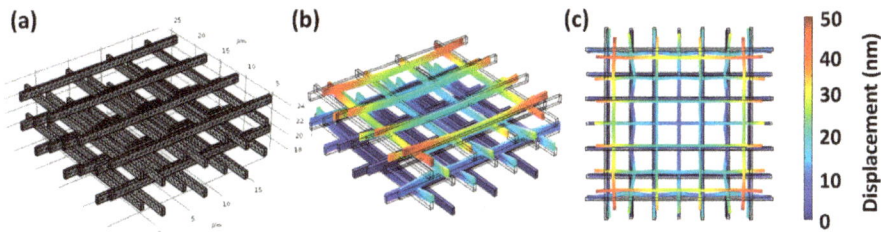

Figure 11. (a) Mesh configuration and (b,c) finite element analysis (FEA) simulation of shrinkage-induced deformation for $24 \times 24 \times 8$ µm³ log-pile block fabricated in IP-Dip with 6.2 µm line spacing. See text for details.

The simulations of the magnitude of rod 3D deformation that is caused by shrinkage (Figure 12c,d) were also in reasonable agreement with the observed behavior in Figures 8 and 10. The main differences between the simulations and observations were for the 100% foam rod. Specifically, the simulation showed a vertical expansion at the rod ends (Figure 12c), which is most likely due to the effect of the material Poisson ratio upon contraction. In the real rods, this did not occur, probably because of the shearing that was observed at the horizontal stitching boundaries, as shown in Figure 10a. In the case of the rod with a cap layer, the simulations compared most closely with the observations. These rods showed reduced shear at the stitching boundaries (Figure 10d) due to the axial constraint of the top layer. Moreover, the simulations correctly predicted the bending of the top cap at the ends of the rod where the shrinkage-induced stresses were the greatest. In this work, we assumed that the observed shrinkage was due to the effects of the unconverted resin, as discussed in Section 3.3. Other earlier studies of shrinkage of 2PP structures reached a similar conclusion [44,46]. However, a quantitative description of shrinkage at the molecular level remains elusive and is a subject of continued interest.

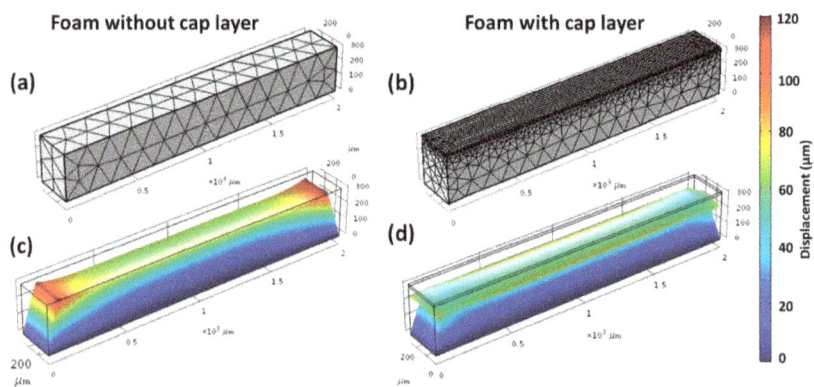

Figure 12. Mesh configuration used for the FEA simulations of foam rods fabricated in IP-Dip resin (a) without and (b) with a 15-µm-thick fully dense cap layer and (c,d) the computed deformation due to shrinkage.

4. Summary and Conclusions

A series of low-atomic number (CHO) millimeter-scale foam laser targets with a 4–6 µm cell size were fabricated using 2PP. The targets are used to support HED physics research, thus driving the requirement that the foam contain only low-atomic number elements. The targets were comprised of a full-density 15 um cap layer at a 0.10 g/cm^3 foam base. The cap layer served as an ablator.

Two commercial acrylic resins were evaluated for preparing the foam targets: (i) IP-Dip, a low viscosity resin designed for high-resolution printing with a large numerical aperture objective and (ii) IP-S a high viscosity resin for mesoscale printing using a lower numerical aperture. A fabricated linewidth in IP-S for different irradiance conditions was reported and compared to prior measurement on IP-Dip and also analyzed using a simple engineering model.

Infrared and Raman spectroscopy were used to measure the extent of 2PP polymerization by monitoring the C=C bond conversion. Although both IP-Dip and IP-S showed significant unconverted material, IP-Dip was worse; the results for IP-Dip agreed with observations that were reported by others. It was proposed that full or partial removal of unreacted resin from within the beams during development was the primary cause of the shrinkage and the resulting deformation observed in the final structures.

Simple end-supported beam test structures were fabricated in each resin to assess the polymer strength and the impact of capillary drying forces. The results showed that simple beams and foams fabricated with these resins were strong enough to support typical capillary drying forces of ~5–10 kPa (~0.7–1.5 psi) without plastic deformation. However, polymer linear shrinkage of up to 6–7% or more during resin development led to large structural plastic deformation in IP-Dip. Finite element analysis was used to simulate the effects of both capillary drying forces and the polymer shrinkage. Drying forces produced elastic deformations <0.5 μm, whereas shrinkage generated ~100× greater axial plastic deformation μm for these target structures.

A significant reduction in shrinkage-induced deformation and improvements in the structure strength and rigidity were achieved by using IP-S resin. Initial tests showed great improvements in the fabricated rod dimensional stability, with up to 4× fewer stitching boundaries and ~1/10th the shrinkage.

Author Contributions: Y.L., J.H.C., O.S. and Y.L. conceived and designed the experiments; Y.L. and O.S. performed the experiments; Y.L., J.H.C., O.S., L.J., J.H., and Y.L. contributed the analysis; Y.L. and J.H.C. wrote the paper. All authors discussed the results and commented on the manuscript at all stages.

Acknowledgments: The authors gratefully acknowledge the financial support of the U.S. Dept. of Energy under contract No. DE-NA0001385. We greatly appreciate the technical support of our many colleagues at the University of Nebraska-Lincoln Laser Assisted Nano-Engineering Laboratory and at Schafer Livermore Laboratory.

Conflicts of Interest: The authors declare no conflict of interest.

References

1. National Research Council. *Frontiers in High Energy Density Physics: The x-Games of Contemporary Science*, 1st ed.; The National Academies Press: Washington, DC, USA, 2003.
2. McCrory, R.; Meyerhofer, D.; Betti, R.; Craxton, R.; Delettrez, J.; Edgell, D.; Glebov, V.Y.; Goncharov, V.; Harding, D.; Jacobs-Perkins, D.; et al. Progress in direct-drive inertial confinement fusion. *Phys. Plasmas* **2008**, *15*, 055503. [CrossRef]
3. Rambo, P.K.; Smith, I.C.; Porter, J.L.; Hurst, M.J.; Speas, C.S.; Adams, R.G.; Garcia, A.J.; Dawson, E.; Thurston, B.D.; Wakefield, C. Z-Beamlet: A multikilojoule, terawatt-class laser system. *Appl. Opt.* **2005**, *44*, 2421–2430. [CrossRef] [PubMed]
4. Spaeth, M.L.; Manes, K.; Kalantar, D.; Miller, P.; Heebner, J.; Bliss, E.; Spec, D.; Parham, T.; Whitman, P.; Wegner, P. Description of the NIF laser. *Fusion Sci. Technol.* **2016**, *69*, 25–145. [CrossRef]
5. Bernat, T.; Campbell, J.; Petta, N.; Sakellari, I.; Koo, S.; Yoo, J.-H.; Grigoropoulos, C. Fabrication of micron-scale cylindrical tubes by two-photon polymerization. *Fusion Sci. Technol.* **2016**, *70*, 310–315. [CrossRef]
6. Jiang, L.; Campbell, J.; Lu, Y.; Bernat, T.; Petta, N. Direct writing target structures by two-photon polymerization. *Fusion Sci. Technol.* **2016**, *70*, 295–309. [CrossRef]
7. Jiang, L.; Campbell, J.; Lu, Y.; Bernat, T.; Petta, N. Precision fabrication of laser targets: Development of 2-photon polymerization as a next-generation tool. In Proceedings of the International Congress on Applications of Lasers & Electro-Optics, Atlanta, GA, USA, 18–22 October 2015.
8. Jiang, L.J.; Maruo, S.; Osellame, R.; Xiong, W.; Campbell, J.H.; Lu, Y.F. Femtosecond laser direct writing in transparent materials based on nonlinear absorption. *MRS Bull.* **2016**, *41*, 975–983. [CrossRef]
9. Stein, O.; Liu, Y.; Streit, J.; Campbell, J.; Lu, Y.; Aglitskiy, Y.; Petta, N. Fabrication of low-density shock-propagation targets using two-photon polymerization. *Fusion Sci. Technol.* **2018**, *73*, 153–165. [CrossRef]
10. Oakdale, J.S.; Smith, R.F.; Forien, J.B.; Smith, W.L.; Ali, S.J.; Bayu Aji, L.B.; Willey, T.M.; Ye, J.; van Buuren, A.W.; Worthington, M.A. Direct laser writing of low-density interdigitated foams for plasma drive shaping. *Adv. Funct. Mater.* **2017**, *27*, 1702425. [CrossRef]
11. Baldacchini, T. *Three-Dimensional Microfabrication Using Two-Photon Polymerization: Fundamentals, Technology, and Applications*, 1st ed.; William Andrew: Waltham, MA. USA, 2015.
12. Lindl, J. Inertial confinement fusion: The quest for ignition and energy gain using indrect drive. *Nucl. Fusion* **1999**, *39*, 825.

13. Mačiulaitis, J.; Deveikytė, M.; Rekštytė, S.; Bratchikov, M.; Darinskas, A.; Šimbelytė, A.; Daunoras, G.; Laurinavičienė, A.; Laurinavičius, A.; Gudas, R. Preclinical study of SZ2080 material 3D microstructured scaffolds for cartilage tissue engineering made by femtosecond direct laser writing lithography. *Biofabrication* **2015**, *7*, 015015. [CrossRef] [PubMed]

14. Jiang, L.; Xiong, W.; Zhou, Y.; Liu, Y.; Huang, X.; Li, D.; Baldacchini, T.; Jiang, L.; Lu, Y. Performance comparison of acrylic and thiol-acrylic resins in two-photon polymerization. *Opt. Express* **2016**, *24*, 13687–13701. [CrossRef] [PubMed]

15. Žukauskas, A.; Matulaitienė, I.; Paipulas, D.; Niaura, G.; Malinauskas, M.; Gadonas, R. Tuning the refractive index in 3D direct laser writing lithography: Towards grin microoptics. *Laser Photonics Rev.* **2015**, *9*, 706–712. [CrossRef]

16. Lemma, E.D.; Rizzi, F.; Dattoma, T.; Spagnolo, B.; Sileo, L.; Qualtieri, A.; De Vittorio, M.; Pisanello, F. Mechanical properties tunability of three-dimensional polymeric structures in two-photon lithography. *IEEE Trans. Nanotechnol.* **2017**, *16*, 23–31. [CrossRef]

17. König, K.; Uchugonova, A.; Straub, M.; Zhang, H.; Licht, M.; Afshar, M.; Feili, D.; Seidel, H. Sub-100 nm material processing and imaging with a sub-15 femtosecond laser scanning microscope. *J. Laser Appl.* **2012**, *24*, 042009. [CrossRef]

18. Oakdale, J.S.; Ye, J.; Smith, W.L.; Biener, J. Post-print uv curing method for improving the mechanical properties of prototypes derived from two-photon lithography. *Opt. Express* **2016**, *24*, 27077–27086. [CrossRef] [PubMed]

19. O'Brien, A.K.; Cramer, N.B.; Bowman, C.N. Oxygen inhibition in thiol-acrylate photopolymerizations. *J. Polym. Sci. Pol. Chem.* **2006**, *44*, 2007–2014. [CrossRef]

20. Jiang, L.J.; Zhou, Y.S.; Xiong, W.; Gao, Y.; Huang, X.; Jiang, L.; Baldacchini, T.; Silvain, J.-F.; Lu, Y.F. Two-photon polymerization: Investigation of chemical and mechanical properties of resins using raman microspectroscopy. *Opt. Lett.* **2014**, *39*, 3034–3037. [CrossRef] [PubMed]

21. Paz, V.F.; Emons, M.; Obata, K.; Ovsianikov, A.; Peterhänsel, S.; Frenner, K.; Reinhardt, C.; Chichkov, B.; Morgner, U.; Osten, W. Development of functional sub-100 nm structures with 3D two-photon polymerization technique and optical methods for characterization. *J. Laser Appl.* **2012**, *24*, 042004. [CrossRef]

22. Greaves, G.N.; Greer, A.; Lakes, R.; Rouxel, T. Poisson's ratio and modern materials. *Nat. Mater.* **2011**, *10*, 823. [CrossRef] [PubMed]

23. Ashby, M. The properties of foams and lattices. *Philos. Trans. R. Soc. Lond. A Math. Phys. Eng. Sci.* **2006**, *364*, 15–30. [CrossRef] [PubMed]

24. Hsu, W.-H.; Masim, F.C.P.; Balčytis, A.; Juodkazis, S.; Hatanaka, K. Dynamic position shifts of x-ray emission from a water film induced by a pair of time-delayed femtosecond laser pulses. *Opt. Express* **2017**, *25*, 24109–24118. [CrossRef] [PubMed]

25. Lee, K.S.; Yang, D.Y.; Park, S.H.; Kim, R.H. Recent developments in the use of two-photon polymerization in precise 2D and 3D microfabrications. *Polym. Adv. Technol.* **2006**, *17*, 72–82. [CrossRef]

26. Serbin, J.; Egbert, A.; Ostendorf, A.; Chichkov, B.; Houbertz, R.; Domann, G.; Schulz, J.; Cronauer, C.; Fröhlich, L.; Popall, M. Femtosecond laser-induced two-photon polymerization of inorganic–organic hybrid materials for applications in photonics. *Opt. Lett.* **2003**, *28*, 301–303. [CrossRef] [PubMed]

27. Leatherdale, C.A.; DeVoe, R.J. Two-photon microfabrication using two-component photoinitiation systems: Effect of photosensitizer and acceptor concentrations, In Nonlinear Optical Transmission and Multiphoton Processes in Organics. In Proceedings of the Optical Science and Technology, Spie's 48th Annual Meeting, San Diego, CA, USA, 3–8 August 2003.

28. Thiel, M.; Fischer, J.; von Freymann, G.; Wegener, M. Direct laser writing of three-dimensional submicron structures using a continuous-wave laser at 532 nm. *Appl. Phys. Lett.* **2010**, *97*, 221102. [CrossRef]

29. Takada, K.; Kaneko, K.; Li, Y.-D.; Kawata, S.; Chen, Q.-D.; Sun, H.-B. Temperature effects on pinpoint photopolymerization and polymerized micronanostructures. *Appl. Phys. Lett.* **2008**, *92*, 041902. [CrossRef]

30. Rekštytė, S.; Jonavičius, T.; Gailevičius, D.; Malinauskas, M.; Mizeikis, V.; Gamaly, E.G.; Juodkazis, S. Nanoscale precision of 3D polymerization via polarization control. *Adv. Opt. Mater* **2016**, *4*, 1209–1214. [CrossRef]

31. Mueller, J.B.; Fischer, J.; Mange, Y.J.; Nann, T.; Wegener, M. In-situ local temperature measurement during three-dimensional direct laser writing. *Appl. Phys. Lett.* **2013**, *103*, 123107. [CrossRef]

32. Juodkazis, S.; Mizeikis, V.; Seet, K.K.; Misawa, H.; Wegst, U.G. Mechanical properties and tuning of three-dimensional polymeric photonic crystals. *Appl. Phys. Lett.* **2007**, *91*, 241904. [CrossRef]
33. Liu, J.-L.; Feng, X.-Q.; Xia, R.; Zhao, H.-P. Hierarchical capillary adhesion of microcantilevers or hairs. *J. Phys. D: Appl. Phys.* **2007**, *40*, 5564. [CrossRef]
34. Yoshimoto, K.; Stoykovich, M.P.; Cao, H.; de Pablo, J.J.; Nealey, P.F.; Drugan, W.J. A two-dimensional model of the deformation of photoresist structures using elastoplastic polymer properties. *J. Appl. Phys.* **2004**, *96*, 1857–1865. [CrossRef]
35. Mastrangelo, C.; Hsu, C. Mechanical stability and adhesion of microstructures under capillary forces. Ii. Experiments. *J. Microelectromech. Syst.* **1993**, *2*, 44–55. [CrossRef]
36. Mastrangelo, C. Adhesion-related failure mechanisms in micromechanical devices. *Tribol. Lett.* **1997**, *3*, 223–238. [CrossRef]
37. Guney, M.; Fedder, G. Estimation of line dimensions in 3D direct laser writing lithography. *J. Micromech. Microeng.* **2016**, *26*, 105011. [CrossRef]
38. Shukla, A.K.; Palani, I.; Manivannan, A. Laser-assisted dry, wet texturing and phase transformation of flexible polyethylene terephthalate substrate revealed by raman and ultraviolet-visible spectroscopic studies. *J. Laser Appl.* **2018**, *30*, 022008. [CrossRef]
39. Suzuki, T.; Morikawa, J.; Hashimoto, T.; Buividas, R.; Gervinskas, G.; Paipulas, D.; Malinauskas, M.; Mizeikis, V.; Juodkazis, S. Thermal and optical properties of sol-gel and su-8 resists. In *Advanced Fabrication Technol. Micro/Nano Optics Photonics V, 82490K*; International Society for Optics and Photonics: Bellingham, WA, USA, 2012.
40. Harris, D.C.; Bertolucci, M.D. *Symmetry and Spectroscopy: An Introduction to Vibrational and Electronic Spectroscopy*; Chapter 3; Dover Publication: New York, NY, USA, 1989.
41. Colthup, N.B.; Daly, L.H.; Wiberley, S.E. *Introduction Infrared Raman Spectroscopy*; Elsevier, Academic Press: San Diego, CA, USA, 2012.
42. Smith, B.C. *Infrared Spectral Interpretation: A Systematic Approach*; CRC Press: Boca Raton, FL, USA, 1998.
43. Socrates, G. *Infrared and Raman Characteristic Group Frequencies: Tables and Charts*; John Wiley & Sons: Chichester, UK, 2001.
44. Ovsianikov, A.; Shizhou, X.; Farsari, M.; Vamvakaki, M.; Fotakis, C.; Chichkov, B.N. Shrinkage of microstructures produced by two-photon polymerization of Zr-based hybrid photosensitive materials. *Opt. Express* **2009**, *17*, 2143–2148. [CrossRef] [PubMed]
45. LaFratta, C.N.; Baldacchini, T. Two-photon polymerization metrology: Characterization methods of mechanisms and microstructures. *Micromachines* **2017**, *8*, 101. [CrossRef]
46. Ovsianikov, A.; Viertl, J.; Chichkov, B.; Oubaha, M.; MacCraith, B.; Sakellari, I.; Giakoumaki, A.; Gray, D.; Vamvakaki, M.; Farsari, M.; et al. Ultra-low shrinkage hybrid photosensitive material for two-photon polymerization microfabrication. *ACS Nano* **2008**, *2*, 2257–2262. [CrossRef] [PubMed]

nanomaterials

MDPI

Article

High Repetition Rate UV versus VIS Picosecond Laser Fabrication of 3D Microfluidic Channels Embedded in Photosensitive Glass

Florin Jipa [1], Stefana Iosub [1], Bogdan Calin [1], Emanuel Axente [1], Felix Sima [1,2,*] and Koji Sugioka [2,*]

[1] Center for Advanced Laser Technologies (CETAL), National Institute for Laser, Plasma and Radiation Physics (INFLPR), 409 Atomistilor, Magurele RO-77125, Romania; florin.jipa@inflpr.ro (F.J.); stefana.iosub@inflpr.ro (S.I.); bogdan.calin@inflpr.ro (B.C.); emanuel.axente@inflpr.ro (E.A.)
[2] RIKEN Center for Advanced Photonics, 2-1 Hirosawa, Wako, Saitama 351-0198, Japan
* Correspondence: felix.sima@inflpr.ro (F.S.); ksugioka@riken.jp (K.S.); Tel.: +4021-4574-491 (F.S.)

Received: 6 July 2018; Accepted: 26 July 2018; Published: 31 July 2018

Abstract: Glass is an alternative solution to polymer for the fabrication of three-dimensional (3D) microfluidic biochips. Femtosecond (fs) lasers are nowadays the most promising tools for transparent glass processing. Specifically, the multiphoton process induced by fs pulses enables fabrication of embedded 3D channels with high precision. The subtractive fabrication process creating 3D hollow structures in glass, known as fs laser-assisted etching (FLAE), is based on selective removal of the laser-modified regions by successive chemical etching in diluted hydrofluoric acid solutions. In this work we demonstrate the possibility to generate embedded hollow channels in photosensitive Foturan glass volume by high repetition rate picosecond (ps) laser-assisted etching (PLAE). In particular, the influence of the critical irradiation doses and etching rates are discussed in comparison of two different wavelengths of ultraviolet (355 nm) and visible (532 nm) ranges. Fast and controlled fabrication of a basic structure composed of an embedded micro-channel connected with two open reservoirs, commonly used in the biochip design, are achieved inside glass. Distinct advantages such as good aspect-ratio, reduced processing time for large areas, and lower fabrication cost are evidenced.

Keywords: picosecond laser processing; 3D microfluidic channels; photosensitive glass

1. Introduction

Microfluidic systems typically consisting of three-dimensionally (3D) embedded channels connected to open micro-reservoirs are useful tools for many biological and medical studies, since they are basic elements for biochips such as lab-on-a-chip devices and micro-total-analysis-systems that can perform reaction, detection, analysis, separation, and synthesis of biochemical materials with high-efficiency, high-speed, high-sensitivity, low reagent consumption, and low waste production [1,2]. The unique 3D geometries offer flexibilities and specific functionalities for fabrication of the biochips [3,4] or even organs-on-a-chip systems [5]. Such microfluidic devices for biomedical applications are generally fabricated based on PDMS with casting technologies [6,7]. However, although they exhibit indubitable advantages such as biocompatibility, good optical quality, and easy to use, some drawbacks including non-reusability, and adsorption of organic compounds, as well as the requirement of multiple stacking and sealing processes in the fabrication procedure push us to find an alternative [8]. During the last decade, femtosecond (fs) laser fabrication has proven to be a powerful tool for 3D internal modification of transparent glass materials and fabrication of embedded channels [9–12]. Thus, glass is a good alternative to PDMS for specific biological applications, which allow creating robust, easy to clean and reusable devices. Among many glasses, Foturan is one of the most suitable materials for fabrication of microfluidic systems. Foturan is a photosensitive glass with photoreactive properties due to the addition of a photoactive agent (photo-sensitizer) and metal ions (nucleation agent) in the glass matrix. The photoactive agent is cerium (<0.04 wt% Ce_2O_3), while the nucleation ion is silver (0.05–0.15 wt% Ag_2O) [13]. Unlike other transparent glasses these agents allow this glass to be processed in a 3D manner by space selective control of the precipitation process [14]. Photoactivation takes place at wavelengths shorter than 350 nm, and then a successive annealing treatment induces silver clustering that converts a latent image of irradiated region into an observable one. During the thermal treatment, a metasilicate crystalline phase is grown around the formed silver clusters which, by an isotropic chemical etching, can be selectively removed to create 3D hollow micropatterns in the glass matrix [15]. Masuda et al. used a high-intensity fs laser emitting light of 150 fs pulse width at 775 nm wavelength, 1 kHz repetition rate and 0.4 μJ, to fabricate complex 3D microfluidic structures inside Foturan glass with a high spatial resolution [16]. In this case, the photo-reaction of near-infrared fs pulses with glass takes place by two-step excitation of electrons with three photon absorptions each, resulting in a six-photon process [17].

This technique enabling the fabrication of 3D microfluidic structures has been termed femtosecond laser-assisted etching (FLAE). FLAE of Foturan glass was then applied to fabricate specific biochips called nano-aquariums for monitoring of continuous motion of *Euglena gracilis* [18] and evaluation of gliding mechanism of *Phormidium* cyanobacteria [19]. Meanwhile, picosecond (ps) lasers, which are also categorized as ultrafast lasers, are becoming more common tools for practical use due to higher power and higher reliability as compared to fs lasers. At this time-scale, the deposition of laser energy is still typically faster than the electron-phonon coupling processes (which are material-dependent), enabling the minimization of heat-affected zone and, thus, high-quality micro- and nano-fabrication [20]. In addition, the high peak power ($P_{peak} = E/\tau$, E—pulse energy, τ—pulse duration) can induce nonlinear absorption processes in materials which do not absorb the laser wavelength, thus allowing the processing not only of the surface, but also of the inside of transparent materials similarly to the fs lasers [9,10]. Therefore, the ps laser may be able to replace the fs lasers for 3D microfabrication of Foturan glass, which is beneficial in terms of the high throughput process.

Veiko et al. have demonstrated the possibility to modify Foturan glass by a ps Nd:YAG laser with a pulse width of 30 ps at 532 nm wavelength and a repetition rate of 10 Hz [21]. A mechanism based on two-photon absorption was proposed. Thus, microstructures attributed to the phase transition from amorphous to crystalline inside the Foturan glass-ceramic material were fabricated by means of local laser modification followed by subsequent thermal treatment. It has been further shown that using the same laser source it is possible to create 3D channels in bulk by hydrofluoric acid (HF) etching of crystallized regions developed by thermal treatment using a CO_2 laser [22]. Even though the proof of concept was demonstrated, the potential of ps lasers for the structuring and fabrication of 3D embedded channels in Foturan glass is still not fully explored and many challenges remained unmet.

In this study we demonstrate the successful fabrication of 3D microfluidic structures in Foturan glass by high repetition rate ps laser-assisted etching (PLAE) using either the second or third harmonics (visible (VIS) 532 nm or ultraviolet (UV) 355 nm wavelengths) of a Nd:YVO$_4$ laser. Critical irradiation doses and etching ratios were examined for both wavelengths to optimize the experimental conditions of PLAE. Controlled fabrication of the microfluidic structures consisting of an embedded channel connected with two open micro-reservoirs is achieved for both cases. Our study evidenced that transparent materials processing with high repetition rate ps laser pulses based on multi-photon absorption could be a viable alternative to classical fs micro-fabrication. More importantly, large-area 3D micro/nanofabrication with considerably reduced processing time and production costs will offer great advantages for manufacturing with high repetition rate ps laser pulses.

2. Materials and Methods

Foturan used in this study is a photo-structurable glass ceramic manufactured by Schott North America Inc., Corporate Office, Elmsford, NY, USA. It is a photo-sensitive alkali-aluminosilicate glass material consisting of SiO_2 (75%–85%), Li_2O (7%–11%), K_2O and Al_2O_3 (3%–6%), Na_2O (1%–2%), ZnO (0%–2%), Sb_2O_3 (0.2%–0.4%), Ag_2O (0.05%–0.15%), and Ce_2O_3 (0.01%–0.04%). In our experiments, we cut Foturan glass wafers to $10 \times 10 \times 2$ mm^3 dimensions to create 3D microfluidic structures. Prior to use, the samples were successively cleaned in baths of acetone, alcohol, and deionized water.

The laser direct writing was conducted by a ps laser beam (Lumera, Coherent) delivering pulses of duration below 10 ps at 500 kHz repetition rate, coupled with a customized workstation (Figure 1). The second (532 nm—VIS) and the third (355 nm—UV) harmonics of a Nd:YVO$_4$ laser were used in experiments with laser power ranging from tens of mW for critical doses up to 500 mW for material processing using VIS wavelength and from values below 1 mW for critical doses up to 10 mW for structuring in UV. For both wavelengths, the scanning speed was varied from 0.1 mm/s to 1 mm/s in order to determine the optimum value of interline spaces which is an important parameter for a line-by-line scanning process. The beam was focused using an aspheric lens of 15 mm focal length producing a circular focal spot of 4 μm in diameter. The entire writing process was monitored with a CCD camera, which was also used to control the focusing position on the sample surface and inside the volume. The samples were placed on an X-Y motorized translation stage (PlanarDL, Aerotech Inc., Pittsburgh, PA, USA) with computer-control which provided 200 mm travel range on each axis with ±500 nm step accuracy and ±100 nm precision. In addition, the stage has the ability to trigger the laser firing for precise synchronization, increasing thus the accuracy of the imprinted pattern.

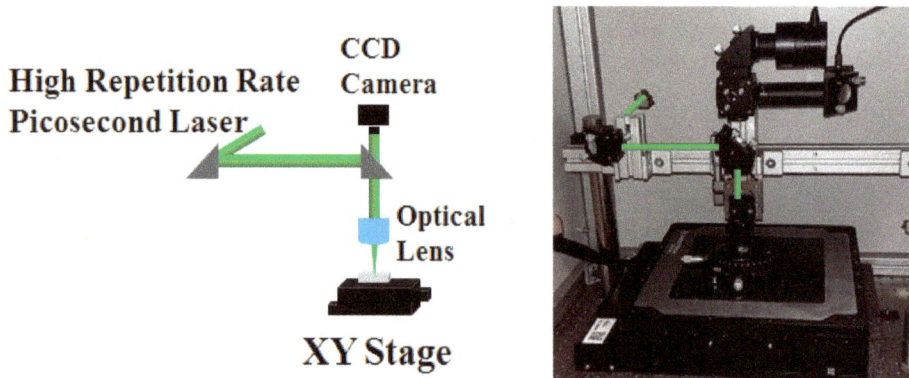

Figure 1. Scheme and photo of the workstation used for high repetition rate ps laser irradiation of Foturan glass.

The exposed samples were then annealed in a furnace (model MTF M1238-250 from Carbolite Gero Limited, Hope Valley, UK) controlled with the following program: heating with a slope of 5 °C/min up to 500 °C, then keeping the temperature constant for 1 h in order to grow Ag nanoparticles, increasing the temperature again with a slope of 3 °C/min to 605 °C, and keeping the temperature constant for another hour to obtain the crystalline phase of Li metasilicate. The process was followed by chemical etching in 8% HF solution under ultrasonic condition. During the etching, the crystalline phase grown around Ag nanoparticles was selectively removed. Profilometry analysis was carried out with a stylus profiler XP2 from Ambios Technology, 0.01 mm/s speed and 1 mm working distance. Optical interrogation of the samples was performed in transmission mode with a microscope, model DM4000 B Led from Leica Microsystems, Wetzlar, Germany. Scanning electron microscopy studies were employed with an FEI Co. microscope, model Inspect S, Hillsboro, OR, USA.

3. Results

3.1. Critical Dose Evaluation

In a first step, the experimental procedure was devoted to the evaluation of the critical dose D_c for each irradiation wavelength, in order to find optimum parameters for laser processing. By using a model proposed by Fuqua et al. [23] as the critical laser fluence F_c, the critical irradiation dose can be expressed as:

$$D_c = F_c^m \times N \tag{1}$$

N is the number of laser pulses necessary to induce photoreaction and Ag nucleation and m represents the number of photons in the multiphoton absorption process for electron generation. F_c is defined as the lowest fluence at which a sufficiently high density of nuclei (Ag nanoclusters) is able to form an interconnected network of the metasilicate crystalline phase in photo-structurable glass by the thermal treatment for selective removal. Here, the density of nuclei ρ generated at a fluence of F can be expressed by $\rho = K \times F^m \times N$ (K is a proportionality constant).

The determination of UV and VIS critical doses was performed by irradiating Foturan glass surface with different laser powers and exposure times. For fixed power values, several exposure times from 0.5 to 3 s were applied, corresponding to 2.5×10^5 to 15×10^5 pulses, resulting in the accumulation of different laser doses in distinct areas. The exposure reproducibility was demonstrated by generating a matrix of identical patterns for each area with a separation distance of 10 μm. This approach allowed easier optical evaluation of the samples and an improved accuracy in D_c evaluation.

For each laser wavelength, several power densities were first applied in order to define the optimal power interval for the D_c evaluation. For a good statistical analysis, we have irradiated 25 identical points using the same laser power and the experiment was repeated twice. No visible differences were observed between experiments. In the case of UV, irradiation the laser power was systematically varied from 0.6 mW to 1 mW with 0.1 mW step (Figure 2a). The same procedure was used for the VIS wavelength for which the laser power was adjusted in the 30–50 mW range with a 5 mW step (Figure 2c).

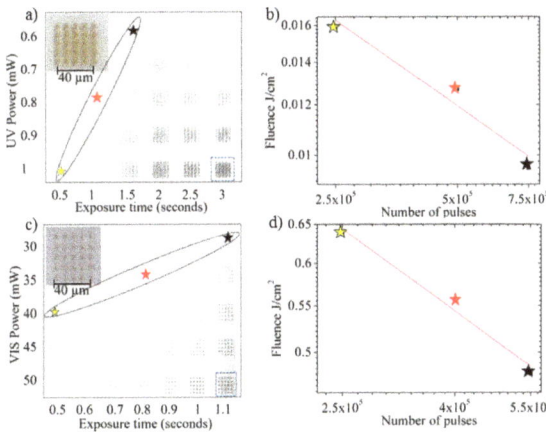

Figure 2. Identification of threshold irradiation parameters for ps laser irradiation of Foturan glass. (**a**,**b**) Optical images of the exposure map written on glass surface using 355 nm ps laser pulses (**a**) and corresponding critical fluence determination (**b**); (**c**,**d**) Optical images of the exposure map written on glass surface using 532 nm ps laser pulses (**c**) and corresponding critical fluence determination (**d**). The inset images represent detailed views of areas marked with square dots.

After irradiation, the glass samples were submitted to a classical thermal treatment protocol described in the previous section. Upon the thermal treatment, brownish crystalline phase became visible on the samples by optical microscopy when the accumulated dose exceeded D_c. The analyses of the irradiated patterns revealed a dose-dependent modification of samples for both laser wavelengths. We identified the irradiation parameters (power as a function of exposure time or number of pulses), at which the glass surface suffered visible modification by laser irradiation followed by thermal treatment (see stars marked in Figure 2a,c) for selective chemical etching, revealing D_c. F_c is a function of the number of pulses for both 355 nm (Figure 2b) and 532 nm (Figure 2d) laser wavelengths. The fluences were estimated for each laser power considering the spot size of 4 μm in diameter. By linearly fitting the log-log representation data, in Figure 2b,d we determined the critical fluences of 1.742 J/cm^2 (m = 3) for VIS and 0.635 J/cm^2 (m = 2) for UV. By introducing these values into Equation (1), the critical dose values for both wavelengths were determined to be D_c = 5.28 J^3/cm^6 for VIS and D_c = 0.4032 J^2/cm^4 for UV, respectively.

The evaluation of D_c is essential in order to further correlate with the translation stage programming for optimizing writing speeds of various complex 3D shapes. In addition, the subsequent wet chemical etching for the subtractive process is critically dependent on the laser irradiation dose for the high precision fabrication of embedded microfluidic channels.

3.2. Etching Rate Estimation

Using the estimated critical doses, we further determined the etching rate dependence for ps UV and VIS lasers, respectively. In HF solution, the contrast ratio in etching selectivity between the unexposed and the laser exposed Foturan glass was found to be 1:50, and was dependent on both the exposure dose and the HF concentration. This ratio was almost coincident with that determined when using the fs laser [24].

We have determined the etching rate of unexposed Foturan glass starting from an initial thickness of 2 mm after immersion in 48 mL solution of 8% HF concentration. At different time intervals, the glass was extracted from solution to measure its thickness with an electronic micrometer for up to 1 h of total exposure. The dependence of the glass thickness on the etching time is plotted in Figure 3a. The evaluation of the linear fitting evidenced an etching rate of about 1.6 μm/min which is in line with previously reported values, such as the study by Helvajian et al. (1.3 μm/min for a slightly lower concentration of 5% HF solution) [25].

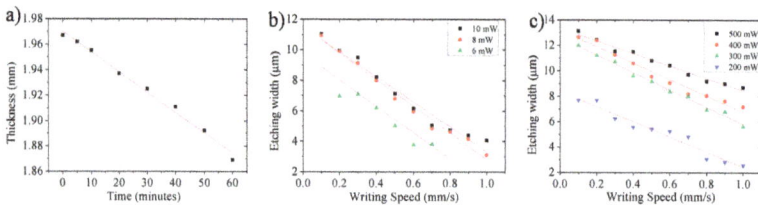

Figure 3. Etching of Foturan glass in 8% HF solution: (**a**) etched thickness as a function of etching time for unexposed glass; (**b**,**c**) width of open channels after 5 min of etching of the linear pattern written by UV (**b**) and VIS (**c**) lasers as a function of writing speed. Each linear regression curve corresponds to different laser power.

The etching rates for VIS and UV lasers exposed samples were estimated by evaluating depths obtained after etching for different time intervals. In both cases, at moderate laser exposures (just above the modification thresholds), ~10 ± 2 μm/min etching rates were obtained. Consequently, the etching ratios were found to be at 1:10. These ratios are dependent on the laser exposure doses. Specifically, an etching rate of 20 μm/min can be achieved when increasing the laser power, corresponding to an etching ratio of up to 1:25. Comparing to etching ratios reported in other publications, where identical wavelength but different pulse widths were used to expose the glass, the values obtained in our study are similar while taking account of processing parameters such as irradiation doses and HF concentration.

The etching characteristics of glass modified by the high repetition rate ps laser were investigated by profilometry measurements. For both laser wavelengths, sets of five identical lines with 1 mm length and 100 µm interspace were written on the glass surface using different irradiation doses. To this purpose, predetermined scanning speeds ranging from 0.1 mm/s to 1 mm/s were employed at different laser powers, followed by annealing treatment. The channel widths created after 5 min of HF etching were evaluated, as presented in Figure 3b,c, for UV and VIS ps laser processing, respectively.

As a general observation, one can notice that the channel widths are dependent on the laser dose, as expected (Figure 3b,c). Indeed, for both wavelengths, a higher dose revealed the formation of wider channels (>10 µm), while only few micrometer widths are obtained for lower doses. In particular, one can obtain 11 µm width channels with UV ps laser pulses of 10 mW power at a writing speed of 0.1 mm/s, and narrower than 4 µm width, with 8 mW average laser power and 1 mm/s writing speed. Similarly, channels with widths wider than 12 µm are obtained by VIS ps laser pulses of 300 mW power and writing speed of 0.1 mm/s, and narrower than 3 µm width, at 200 mW laser power and 1 mm/s writing speed. These results allow us to predict the optimum laser processing parameters and scanning regimes for choosing appropriate interspaces between linear patterns in order to create more complex 3D structures.

3.3. Fabrication of Microfluidic Embedded Channels

Further, the capability of generating embedded microfluidic channels in photosensitive glass using either UV or VIS high repetition rate ps laser pulses was demonstrated. By the new PLAE technique, we propose fabricating a simple microfluidic structure consisting of two open micro-reservoirs connected by an embedded channel (Figure 4). Specifically, square reservoirs of 1×1 mm^2 area are connected by a 1 mm length and 300 µm width channel. The whole structure was designed and fabricated by irradiating the glass in a two-layer configuration (see Figure 4a): (*i*) the first layer (upper part) of the reservoirs (opening) were formed by scanning the focused laser beam on the glass surface and then (*ii*) the second layer (bottom part) of the reservoirs and connecting channel were formed by changing the laser focusing position from the glass surface into the volume at specific depths. Each layer consists of parallel lines written by linear scanning of the focused laser beam at a speed of 0.5 mm/s with lateral sliding at a 5 µm step between each line. The PLAE process was then followed by a two-step treatment at 500 °C for one hour followed by one at 605 °C for another hour (Figure 4b) and successive chemical etching in HF (Figure 4c) in order to obtain the designed structure. The etching time depends on the exposure parameters, which were varied between 50 and 60 min, similar with times used for FLAE.

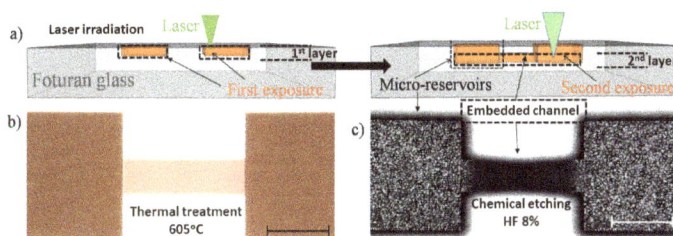

Figure 4. PLAE process of Foturan glass. (**a**) Sketch of the proposed design (with two open reservoirs connected by an embedded channel) and laser irradiation using two-layer configuration; and (**b,c**) partial view by optical microscopy of the obtained structure after irradiation with the VIS laser followed by (**b**) thermal treatment and (**c**) chemical etching in 8% HF solution. The scale bar represents 0.5 mm.

We created two structures written with the identical writing scheme using VIS-PLAE at 500 mW laser power, 1 µJ energy, and 10 µm interline spaces, but at different scanning speeds to investigate influence of the exposure dose on the created structures. The laser irradiation was carried out on the surface (first layer) and at 250 µm depth (second layer) with writing speeds fixed at 0.1 mm/s (Figure 5a,b) and 0.9 mm/s (Figure 5c) respectively. SEM analyses have revealed that a lower irradiation dose (correlated with faster writing speed and smaller number of applied laser pulses) can generate embedded microfluidic channels in glass volume after 50 min of wet chemical etching (Figure 5b,c). Contrarily, when slower speed (higher irradiation dose) is applied, the channel rooftop disappeared due to second layer overexposure (Figure 5a). Indeed, by increasing the irradiation dose inside glass, the volume affected by multiphoton absorption becomes larger. Consequently, if the position of the second layer is close enough to the surface, the channel rooftop can be removed during the chemical etching. It is worth mentioning that fabrication of the structure obtained at the scanning speed of 0.1 mm/s (Figure 5a) required a processing time of about 70 min, while the structure fabricated at 0.9 mm/s (Figure 5c) could be written in only 10 min. One may further adjust the thickness of channel rooftop by tailoring the scanning speed.

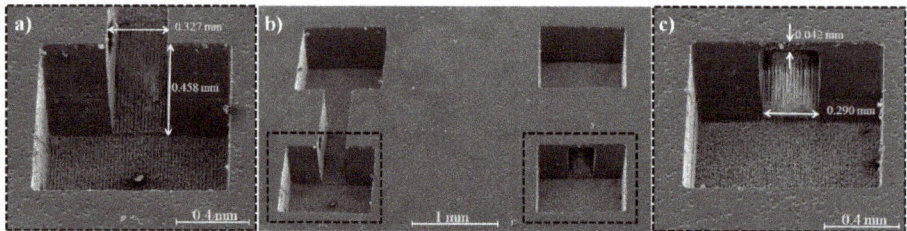

Figure 5. SEM images of structures fabricated by VIS-PLAE using a ps laser at power of 500 mW and a design of 10 µm interline spaces, at two different scanning speeds: 0.1 mm/s (**a,b**) and 0.9 mm/s (**b,c**). The laser irradiation was carried out with two-layer configuration: on the surface (first layer) and at 250 µm depth (second layer).

Fabrication of the similar structure was also attempted with the same writing scheme by ps laser processing at a wavelength of 355 nm. After finding optimal parameters, embedded channels could be also fabricated by PLAE using a UV ps laser (UV-PLAE). In particular, by applying a power of 12 mW, energy of 0.024 µJ, a 5 µm interline step and a writing speed of 0.5 mm/s the structure with two open reservoirs connected by embedded channel was successfully created. With these processing conditions, the entire structure was written in less than 20 min and an etching time of 60 min.

In Figure 6, we present SEM images of twin structures separated by 2 mm written in Foturan glass by UV-PLAE. Pairs of two open square reservoirs of 1×1 mm^2 area are connected by 1 mm length and 335 µm width embedded channels.

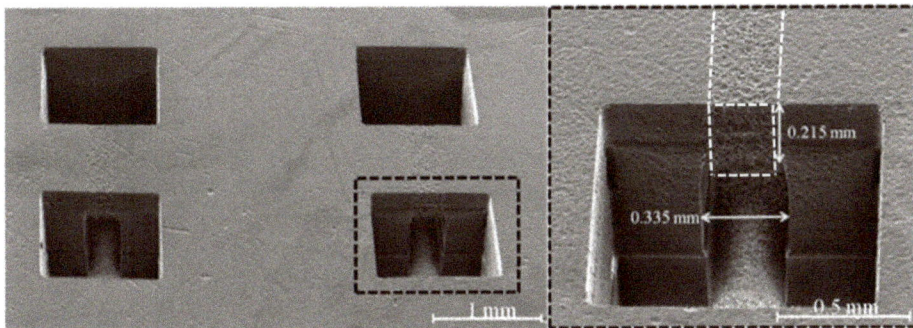

Figure 6. SEM images of twin structures fabricated by UV-PLAE using a ps laser of 12 mW laser power and 5 μm interline spaces, at 0.5 mm/s scanning speed. The laser irradiation was carried out using two-layer configuration: (on the surface (first layer) and at 500 μm depth (second layer)).

4. Discussion

During interaction with solid materials, the pulsed laser beam is depositing its energy, inducing different phenomena dependent on the pulse energy, duration, and focusing optics [26]. It is rather a thermal process for long pulses (nanosecond regime), while a physical aspect predominates for ultrashort pulses (less than a few ps pulses that are shorter than electron-phonon coupling time in materials) [27]. Consequently, in the case of ultrashort pulses, the heat-affected zone is minimized [20,28,29]. In most of the studies, the photo-physical and photo-chemical mechanisms involved during glass processing with ultrashort pulses are evidenced for femtosecond laser pulses and extensively addressed, from both fundamental and applicative points of view [12,30–32]. Origins of material modification under interaction with fs pulses could be related to the densification induced by pressure wave and/or fast heating-cooling processes [33,34]. As a result, electrons in the conduction band are heated by the laser pulse very quickly so that they do not have time to diffuse out from the irradiated volume or to recombine. The photoionization is then responsible of seeding electrons for the subsequent avalanche ionization. It was found out that the electron density increases by avalanche ionization until its plasma frequency reaches the critical plasma density [35]. On the other hand, the avalanche ionization is more efficient for longer pulses since it allows more time for increasing the electron density. In our study, ps laser irradiance values of 3.73×10^{13} and of 3.57×10^{12} Wcm^{-2} for VIS and UV, respectively, were determined, below values of 10^{14} Wcm^{-2} for fs laser irradiances [16]. These values for ps laser irradiances corresponds to a Keldysh parameter above 1.5 which indicates the multi-photon absorption is more dominant rather than tunneling ionization in the case of both UV and VIS irradiation [36]. In contrast, pulses longer than a few tens of ps do not reach enough intensity to directly photoionize the electrons.

In the first studies using ultrafast laser reported in literature, Foturan photosensitive glasses were exposed to 150 fs laser pulses at 775 nm wavelength [16]. A second thermal treatment step followed by isotropic etching conducted to a preferential material removal from irradiated regions. In this case, the experimental investigation of F_c on the number of pulses allowed the calculation of a critical dose of $D_c = 1.3 \times 10^{-5}$ J^6/cm^{12} with $m = 6$ for the 775 nm wavelength fs laser. Interaction mechanisms of laser pulses with Foturan glass were further explored by employing: (*i*) fs laser (150 fs, 775 nm, 1 kHz), and ns lasers of (*ii*) 266 nm; (*iii*) 355 nm; and (*iv*) 308 nm laser wavelengths [17]. A significant increase in absorption spectrum of the exposed samples around 360 nm was found for fs irradiated glasses corresponding to absorption from oxygen deficient centers which originated from the interband excitation of electrons. The absorption at 315 nm related to Ce ions was not observed, suggesting that Ce^{3+} ions do not contribute to electron generation for the reduction to Ag atoms in the case of fs-irradiated samples at 775 nm. Thus, it was concluded that free electrons were generated by two-step interband excitation through the defect levels with three-photon absorption each resulting in six photons in total for photoreaction. A similar photoreaction mechanism by successive interband excitation was evidenced for 266 nm ns laser-irradiated samples, but by a linear two-photon process, i.e., two-step excitation by single-photon absorption each. On the other hand, in the case of 355 nm laser-irradiated samples, free electrons are generated by Ce^{3+}, while in the case of the 308 nm laser both absorption by Ce^{3+} (single-photon absorption) and interband excitation (the linear two-photon process) was found. In this study, we evaluated the possibility to obtain embedded microfluidic channels in Foturan glass by using a high repetition rate (500 kHz) laser of <10 ps pulse duration at 532 nm and 355 nm wavelengths. A nonlinear absorption process is evident at the interaction of both 532 nm and 355 nm laser pulses with Foturan glass, since the absorption edge of Foturan is shorter than 350 nm. The Foturan glass also has an absorption peak around 315 nm ascribed to Ce^{3+} absorption, which corresponds to photon energy of 3.93 eV. Therefore, for 532 nm with the photon energy of 2.33 eV, two photons are necessary to generate free electrons from Ce^{3+}. Another channel for free electron generation is interband excitation, for which the two-step excitation model through the intermediate state has been proposed [17]. For this excitation, the photon energy of 3.49–4.66 eV is required for each step, indicating four photons in total for the 532 nm beam. We have found $m = 3$ photon process for VIS pulses. Thus, it is likely that the free electrons are generated by both the Ce^{3+} absorption and the interband excitation through the intermediate state similarly to the case of the 308 nm ns laser [17]. In the case of UV pulses, at 355 nm, with the photon energy of 3.49 eV, we found $m = 2$ photon process which should correspond to free electron generation by two-photon absorption by Ce^{3+} similarly to the case of the 355 nm ns laser [17]. In addition, we consider that, in our experimental conditions, the high repetition rate of the ps laser pulses can increase the nonlinear absorptivity into Foturan glass material [37]. The alteration of the surface and volume morphologies that can be observed above the embedded channel in Figure 6 (white dotted regions) may support the hypothesis that, for high repetition rate UV ps laser pulses, the heat effect becomes more important when the laser beam is focused above a critical intensity.

During interaction with a Gaussian laser beam (fs or ps durations) with Foturan glass, the absorption profile affects the shape of modification volume to develop an elliptical cross-sectional shape of the crystallized area. In case of a high intensity fs laser pulse, the main advantage resides in high density, compact modification at very low power, which can be used for high resolution micro-processing. On the other hand, ps lasers can represent a viable alternative for large-scale processing both on surface and in volume due to high average power compensation since longer, high-energy, pulses can modify more material per pulse. Indeed, a larger energy can be deposited in the material since the time is longer. Thus, this energy allows more time for growth of electron density during the laser irradiation and in consequence an increase of formed Ag atoms. These Ag atoms are then responsible of the formation of larger Ag clusters and a larger glass crystalline area during thermal treatment. As a result, one may consider that the area of the crystalline phase is dependent on pulse duration. Using the same focusing optics, ps laser beams can consequently decrease the processing times as compared with femtosecond lasers. As a direct comparison, to create a similar structure, irradiation time was 1.5 h for laser pulses of 360 fs at 522 nm (2 mm/s scanning speed and 250 KHz laser repetition rate) while, in our case, it was of approximately 10 min at a scanning speed of 0.9 mm/s and 500 KHz (SI in [38]). On the other hand, in case of shorter pulses one needs less energy to achieve the intensity for optical breakdown allowing the achievement more precise machining with femtosecond lasers rather than the longer pulse lasers. Thus, higher resolution processing is achieved with fs pulses than ps pulses. In case of ps laser pulses, a more efficient process could be achieved for UV pulses since the critical dose is one order of magnitude less than in case of VIS pulses compensating the laser power conversion efficiency (1:2 in case of VIS and 1:3 for UV).

The high repetition rate ps laser pulses could thus stand as a prospective processing benefit, attractive for several applications that require high speed and cost-effective manufacturing. On the other hand, by controlling the irradiation dose with respect to the laser wavelength and etching parameters one can explore the unique characteristics in 3D microfabrication for a wide range of applications.

We have finally demonstrated that PLAE is a suitable and very fast fabrication method of 3D embedded microfluidic channels with good aspect ratio and sharp edges without any cracks by using either ps laser pulses in UV or VIS.

5. Conclusions

High repetition rate ps laser processing at both 532 and 355 nm wavelengths was applied for fabrication of 3D microfluidic structures in Foturan glass. A three-photon process with a critical fluence of 1.742 J/cm^2 for the VIS case and a two-photon process with 0.635 J/cm^2 for the UV case were found. Critical dose values of 5.28 J^3/cm^6 for VIS and 0.4032 J^2/cm^4 for UV cases were calculated.

Straight lines of 1 mm length and 100 μm interspace were then written on the glass surface at different scanning speeds using different laser powers. Open channels with widths ranging from 3 to 13 μm were developed by thermal treatment and HF etching depending on irradiation doses. Based on a subtractive fabrication process consisting of selective removal of laser-modified regions by chemical etching, we could further fabricate 3D hollow structures in glass by PLAE at both 355 nm and 532 nm laser wavelengths. Due to high power, high repetition rate laser pulses which increase multiphoton absorption, we could apply the laser irradiation process very fast for fabrication of embedded structures. A simple configuration consisting of two micro-reservoirs connected by an embedded channel can be achieved in less than 10 min of laser irradiation by the PLAE technique using either 355 or 532 nm wavelengths.

Author Contributions: Conceptualization: F.S.; methodology: F.J., S.I., and B.C.; investigation: F.J., S.I., and E.A.; writing—original draft preparation: F.S.; writing—review and editing: F.S. and K.S.; project administration: F.S.

Funding: This research was funded by UEFISCDI grants 131PED/2017, 148/PED2017 and TE7/2018.

Acknowledgments: The authors are grateful to Iuliana Iordache for profilometry measurements and to Gianina Popescu-Pelin for SEM support.

Conflicts of Interest: The authors declare no conflict of interest.

References

1. Sackmann, E.K.; Fulton, A.L.; Beebe, D.J. The present and future role of microfluidics in biomedical research. *Nature* **2014**, *507*, 181–189. [CrossRef] [PubMed]
2. Yeo, L.Y.; Chang, H.C.; Chan, P.P.; Friend, J.R. Microfluidic devices for bioapplications. *Small* **2011**, *7*, 12–48. [CrossRef] [PubMed]
3. Sima, F.; Serien, D.; Wu, D.; Xu, J.; Kawano, H.; Midorikawa, K.; Sugioka, K. Micro and nano-biomimetic structures for cell migration study fabricated by hybrid subtractive and additive 3D femtosecond laser processing. In Proceedings of the Laser-Based Micro- and Nanoprocessing XI (LBMP), San Francisco, CA, USA, 31 January–2 February 2017; Volume 10092, pp. 1009207–1009215.
4. Sima, F.; Xu, J.; Wu, D.; Sugioka, K. Ultrafast laser fabrication of functional biochips: New avenues for exploring 3D micro-and nano-environments. *Micromachines* **2017**, *8*, 40. [CrossRef]
5. Caplin, J.D.; Granados, N.G.; James, M.R.; Montazami, R.; Hashemi, N. Microfluidic organ-on-a-chip technology for advancement of drug development and toxicology. *Adv. Healthc. Mater.* **2015**, *4*, 1426–1450. [CrossRef] [PubMed]
6. Fujii, T. Pdms-based microfluidic devices for biomedical applications. *Microelectron. Eng.* **2002**, *61*, 907–914. [CrossRef]
7. Sia, S.K.; Whitesides, G.M. Microfluidic devices fabricated in poly (dimethylsiloxane) for biological studies. *Electrophoresis* **2003**, *24*, 3563–3576. [CrossRef] [PubMed]
8. Bhatia, S.N.; Ingber, D.E. Microfluidic organs-on-chips. *Nat. Biotechnol.* **2014**, *32*, 760–772. [CrossRef] [PubMed]

9. Itoh, K.; Watanabe, W.; Nolte, S.; Schaffer, C.B. Ultrafast processes for bulk modification of transparent materials. *MRS Bull.* **2006**, *31*, 620–625. [CrossRef]
10. Gattass, R.R.; Mazur, E. Femtosecond laser micromachining in transparent materials. *Nat. Photonics* **2008**, *2*, 219–225. [CrossRef]
11. Sugioka, K.; Cheng, Y. Femtosecond laser processing for optofluidic fabrication. *Lab Chip* **2012**, *12*, 3576–3589. [CrossRef] [PubMed]
12. Sima, F.; Sugioka, K.; Vázquez, R.M.; Osellame, R.; Kelemen, L.; Ormos, P. Three-dimensional femtosecond laser processing for lab-on-a-chip applications. *Nanophotonics* **2017**, *7*, 613–634. [CrossRef]
13. Livingston, F.; Adams, P.; Helvajian, H. Influence of cerium on the pulsed uv nanosecond laser processing of photostructurable glass ceramic materials. *Appl. Surf. Sci.* **2005**, *247*, 526–536. [CrossRef]
14. Stookey, S. Catalyzed crystallization of glass in theory and practice. *Ind. Eng. Chem.* **1959**, *51*, 805–808. [CrossRef]
15. Livingston, F.; Helvajian, H. Variable UV laser exposure processing of photosensitive glass-ceramics: Maskless micro-to meso-scale structure fabrication. *Appl. Phys. A* **2005**, *81*, 1569–1581. [CrossRef]
16. Masuda, M.; Sugioka, K.; Cheng, Y.; Aoki, N.; Kawachi, M.; Shihoyama, K.; Toyoda, K.; Helvajian, H.; Midorikawa, K. 3D microstructuring inside photosensitive glass by femtosecond laser excitation. *Appl. Phys. A: Mater. Sci. Process.* **2003**, *76*, 857–860. [CrossRef]
17. Hongo, T.; Sugioka, K.; Niino, H.; Cheng, Y.; Masuda, M.; Miyamoto, I.; Takai, H.; Midorikawa, K. Investigation of photoreaction mechanism of photosensitive glass by femtosecond laser. *J. Appl. Phys.* **2005**, *97*, 063517. [CrossRef]
18. Hanada, Y.; Sugioka, K.; Kawano, H.; Ishikawa, I.S.; Miyawaki, A.; Midorikawa, K. Nano-aquarium for dynamic observation of living cells fabricated by femtosecond laser direct writing of photostructurable glass. *Biomed. Microdevices* **2008**, *10*, 403–410. [CrossRef] [PubMed]
19. Hanada, Y.; Sugioka, K.; Shihira-Ishikawa, I.; Kawano, H.; Miyawaki, A.; Midorikawa, K. 3D microfluidic chips with integrated functional microelements fabricated by a femtosecond laser for studying the gliding mechanism of cyanobacteria. *Lab Chip* **2011**, *11*, 2109–2115. [CrossRef] [PubMed]
20. Le Harzic, R.; Huot, N.; Audouard, E.; Jonin, C.; Laporte, P.; Valette, S.; Fraczkiewicz, A.; Fortunier, R. Comparison of heat-affected zones due to nanosecond and femtosecond laser pulses using transmission electronic microscopy. *Appl. Phys. Lett.* **2002**, *80*, 3886–3888. [CrossRef]
21. Ageev, E.; Kieu, K.; Veiko, V.P. Modification of photosensitive glass-ceramic foturan by ultra short laser pulses. In Proceedings of the Fundamentals of Laser Assisted Micro and Nanotechnologies, St Petersburg, Russia, 5–8 July 2010.
22. Sergeev, M.; Veiko, V.; Tiguntseva, E.; Olekhnovich, R. Picosecond laser fabrication of microchannels inside foturan glass at CO_2 laser irradiation and following etching. *Opt. Quantum Electron.* **2016**, *48*, 485. [CrossRef]
23. Fuqua, P.D.; Janson, S.W.; Hansen, W.W.; Helvajian, H. Fabrication of true 3D microstructures in glass/ceramic materials by pulsed uv laser volumetric exposure techniques. In Proceedings of the Laser Applications in Microelectronic and Optoelectronic Manufacturing IV, San Jose, CA, USA, 1999; Volume 3618, pp. 213–221.
24. Sugioka, K.; Cheng, Y. Fabrication of 3d microfluidic structures inside glass by femtosecond laser micromachining. *Appl. Phys. A* **2014**, *114*, 215–221. [CrossRef]
25. Hansen, W.W.; Janson, S.W.; Helvajian, H. Direct-write uv-laser microfabrication of 3D structures in lithium-aluminosilicate glass. In Proceedings of the Laser Applications in Microelectronic and Optoelectronic Manufacturing II, San Jose, CA, USA, 1997; Volume 2991, pp. 104–112.
26. Allmen, M.; Blatter, A. *Laser-Beam Interactions with Materials: Physical Principles and Applications*; Springer Science & Business Media: Berlin/Heidelberg, Germany, 2013; Volume 2.
27. Chichkov, B.N.; Momma, C.; Nolte, S.; Von Alvensleben, F.; Tünnermann, A. Femtosecond, picosecond and nanosecond laser ablation of solids. *Appl. Phys. A* **1996**, *63*, 109–115. [CrossRef]
28. Singh, J.P.; Thakur, S.N. *Laser-Induced Breakdown Spectroscopy*; Elsevier: Amsterdam, The Netherlands, 2007.
29. Harilal, S.S.; Freeman, J.R.; Diwakar, P.K.; Hassanein, A. Femtosecond laser ablation: Fundamentals and applications. In *Laser-Induced Breakdown Spectroscopy*; Springer: Berlin/Heidelberg, Germany, 2014; pp. 143–166.
30. Sugioka, K.; Cheng, Y. Femtosecond laser three-dimensional micro-and nanofabrication. *Appl. Phys. Rev.* **2014**, *1*, 041303. [CrossRef]

31. Malinauskas, M.; Žukauskas, A.; Hasegawa, S.; Hayasaki, Y.; Mizeikis, V.; Buividas, R.; Juodkazis, S. Ultrafast laser processing of materials: From science to industry. *Light Sci. Appl.* **2016**, *5*, e16133. [CrossRef]
32. Jiang, L.J.; Maruo, S.; Osellame, R.; Xiong, W.; Campbell, J.H.; Lu, Y.F. Femtosecond laser direct writing in transparent materials based on nonlinear absorption. *MRS Bull.* **2016**, *41*, 975–983. [CrossRef]
33. Schaffer, C.B.; Brodeur, A.; García, J.F.; Mazur, E. Micromachining bulk glass by use of femtosecond laser pulses with nanojoule energy. *Opt. Lett.* **2001**, *26*, 93–95. [CrossRef] [PubMed]
34. Sakakura, M.; Terazima, M.; Shimotsuma, Y.; Miura, K.; Hirao, K. Observation of pressure wave generated by focusing a femtosecond laser pulse inside a glass. *Opt. Express* **2007**, *15*, 5674–5686. [CrossRef] [PubMed]
35. Wu, A.Q.; Chowdhury, I.H.; Xu, X. Femtosecond laser absorption in fused silica: Numerical and experimental investigation. *Phys. Rev. B* **2005**, *72*, 085128. [CrossRef]
36. Schaffer, C.B.; Brodeur, A.; Mazur, E. Laser-induced breakdown and damage in bulk transparent materials induced by tightly focused femtosecond laser pulses. *Meas. Sci. Technol.* **2001**, *12*, 1784. [CrossRef]
37. Miyamoto, I.; Cvecek, K.; Schmidt, M. Evaluation of nonlinear absorptivity in internal modification of bulk glass by ultrashort laser pulses. *Opt. Express* **2011**, *19*, 10714–10727. [CrossRef] [PubMed]
38. Wu, D.; Wu, S.Z.; Xu, J.; Niu, L.G.; Midorikawa, K.; Sugioka, K. Hybrid femtosecond laser microfabrication to achieve true 3D glass/polymer composite biochips with multiscale features and high performance: The concept of ship-in-a-bottle biochip. *Laser Photonics Rev.* **2014**, *8*, 458–467. [CrossRef]

nanomaterials

MDPI

Article

Femtosecond Laser-Based Modification of PDMS to Electrically Conductive Silicon Carbide

Yasutaka Nakajima [1], Shuichiro Hayashi [2], Akito Katayama [1], Nikolay Nedyalkov [3] and Mitsuhiro Terakawa [1,2,*

[1]   School of Integrated Design Engineering, Keio University, 3-14-1, Hiyoshi, Kohoku-ku, Yokohama 223-8522, Japan; y.nakajima@tera.elec.keio.ac.jp (Y.N.); a.katayama@tera.elec.keio.ac.jp (A.K.)
[2]   Department of Electronics and Electrical Engineering, Keio University, 3-14-1, Hiryoshi, Kohoku-ku, Yokohama 223-8522, Japan; s.hayashi@tera.elec.keio.ac.jp
[3]   Institute of Electronics, Bulgarian Academy of Sciences, Tzarigradsko shouse 72, Sofia 1784, Bulgaria; nned@ie.bas.bg
*   Correspondence: terakawa@elec.keio.ac.jp; Tel.: +81-45-566-1737

Received: 28 June 2018; Accepted: 19 July 2018; Published: 22 July 2018

Abstract: In this paper, we experimentally demonstrate femtosecond laser direct writing of conductive structures on the surface of native polydimethylsiloxane (PDMS). Irradiation of femtosecond laser pulses modified the PDMS to black structures, which exhibit electrical conductivity. Fourier-transform infrared (FTIR) and X-ray diffraction (XRD) results show that the black structures were composed of β-silicon carbide (β-SiC), which can be attributed to the pyrolysis of the PDMS. The electrical conductivity was exhibited in limited laser power and scanning speed conditions. The technique we present enables the spatially selective formation of β-SiC on the surface of native PDMS only by irradiation of femtosecond laser pulses. Furthermore, this technique has the potential to open a novel route to simply fabricate flexible/stretchable MEMS devices with SiC microstructures.

Keywords: femtosecond laser; silicon carbide; polydimethylsiloxane; laser direct writing

---

1. Introduction

Polydimethylsiloxane (PDMS) is a widely used polymer in various applications, including wearable/implantable devices and microfluidics, owing to its biocompatibility, optical transparency, flexibility, and elasticity [1]. Recently, PDMS has attracted considerable attention as a soft material to be utilized for flexible/stretchable electrical devices, such as stretchable displays [2] and stretchable strain sensors [3]. In such flexible/stretchable electrical devices, precise micro- or nano-sized structures composed of electrically conductive materials, e.g., metals or carbon materials, or semiconductor materials are essential. Photolithography has been used for the fabrication of microstructures composed of electrically conductive materials on the surface of or inside PDMS; however, the method requires multiple steps for the fabrication [4–6]. Multi-photon photoreduction of metal ions induced by femtosecond laser pulses enables the fabrication of metal structures on the surface of/or inside a soft material [7,8]. A method based on the photoreduction of a metal ink was also reported to fabricate metal structures on the surface of a soft material [9]. These methods require the additional doping of electrically conductive materials, including metal nanoparticles and metal ions.

It is challenging to form a conductive structure by laser irradiation without doping an additional material to polymers. Carbonization of polymers by laser irradiation enables the direct writing of an electrically conductive structure on polymers [10,11]. Rahim et al. reported the spatially selective fabrication of an electrically conductive carbon structure by the carbonization of a polyimide by laser irradiation [10]. The carbon structures fabricated on the surface of the polyimide were transferred to the surface of PDMS to fabricate a PDMS-based stretchable strain sensor. For direct modification of PDMS, the formation of carbon materials by irradiating 800-nm femtosecond laser pulses [12] or ultraviolet nanosecond laser pulses [13] to PDMS was reported. A limited number of papers reported the formation of semiconductor structures on the surface of or inside PDMS using a laser. By the irradiation of 527-nm or 1064-nm femtosecond laser pulses [12,14] or 532-nm or 1064-nm nanosecond laser pulses [13,14], c-silicon was formed on the surface of PDMS by chemical modification. However, the fabrication of electrically conductive structures on the surface of PDMS by direct modification of PDMS using a laser has not been reported, despite the demands for various PDMS-based electrical devices.

In this study, we present the formation of electrically conductive structures on the surface of PDMS by irradiation with femtosecond laser pulses. The electrical conductivity of the formed structures is measured. Analytical Fourier-transform infrared (FTIR) spectroscopy and X-ray diffraction (XRD) results showed that the formed structures were composed of β-silicon carbide (β-SiC). To the best of our knowledge, this is the first demonstration of the direct modification of native PDMS to a conductive material composed of SiC. The presented method enables a direct fabrication of conductive lines in a biocompatible polymer with potential applications in microelectromechanical systems (MEMS) and electro-bioimplants.

2. Materials and Methods

Liquid photo-curable PDMS (KER-4690A/B, Shin-Etsu Chemical Co., Ltd., Tokyo, Japan) in a mold was illuminated using an ultraviolet lamp at a wavelength of 365 nm for 30 min. The polymerized PDMS was rinsed with ethanol, which prevented the adhesion of PDMS to a cover glass on which PDMS was placed during the laser irradiation. Laser pulses with a central wavelength of 522 nm (the second harmonic wave of a 1045-nm femtosecond laser (High Q-2, Spectra-Physics, Santa Clara, CA, USA)), a pulse duration of 192 fs, and a repetition rate of 63 MHz were used for laser direct writing. Femtosecond laser pulses focused by an objective lens (numerical aperture (NA) of 0.4, Olympus, Tokyo, Japan) were irradiated to the lower surface of the PDMS from the bottom in air. The lower surface of the PDMS had contact with the cover glass. The beam diameter d at the focal point was assumed to be ~1.6 μm, according to the formula $d = 1.22\lambda/NA$. Laser power used for proof-of-concept experiments was 150 mW. Using a xyz-translation stage, the samples were scanned on the xy-plane. In the scanned area, adjacent lines were sufficiently overlapped so that the scanned area was assumed to be fully modified in the area. In experiments for the characterization of structures formed under different irradiation conditions, laser pulses were irradiated to the lower surface of the PDMS with a 140-μm air-gap between the surface of PDMS and the surface of the cover glass, which was performed in order to exclude the effect of the cover glass on the fabrication of structures on the lower surface of the PDMS. For this experiment, laser power was varied from 70 mW to 350 mW. The fabrication process was monitored in real time with a CMOS camera (Thorlabs, Newton, NJ, USA).

The structures formed by laser irradiation were observed by optical microscopy and scanning electron microscopy (SEM, Inspect F50, FEI, Hillsboro, OR, USA). Also, the formed structures were examined with FTIR spectroscopy (ALPHA-E, Bruker, Billerica) and XRD (D8 Discover, Bruker, Billerica, MA, USA). FTIR was performed for the wavenumbers, 400 to 4000 cm^{-1}. XRD was performed for 2θ, 12.4° to 100°. For XRD, a generation voltage of 40 kV was used. Current–voltage curves of the fabricated structures were obtained in the range from 0 to 10 V in 0.1 V steps by two-terminal measurement using a digital source meter (2401, Keithley, Cleveland, OH, USA). Probes were set 6 mm apart from each other on the surface of the fabricated structures for all the experiments. Average resistance was determined by calculating the resistance at each voltage, using the obtained measurements.

3. Results and Discussion

3.1. Formation of Conductive Structures on PDMS by Femtosecond Laser Pulse Irradiation

Multiple line structures were fabricated with a line–to–line interval of 20 μm by focused femtosecond laser pulses at a scanning speed of 2 mm/s in the *x*-direction. The lengths of the scanned area were 5 mm in both *x*- and *y*-directions. A photographic image of the structure fabricated on the surface of PDMS is shown in Figure 1a. The irradiated surface changed from optically transparent to black-colored; however, no obvious laser ablation was visually observed. Microscale surface roughness was observed on the black structure with SEM (Figure 1b). The direction of the observed ripple structures corresponded to the scanning direction. The line-to-line interval, i.e., the scanning interval, was 25 μm, which was comparable to the period of the formed ripple structures. The peak laser intensity at the focal point under the experimental condition is estimated to be 6.2×10^{11} J/cm^2, which is lower than the estimated peak intensity used for laser ablation of PDMS under the conditions of a laser wavelength of 527 nm, pulse duration of 300 fs, and repetition rate of 33 Hz [12]. We performed the experiment at a repetition rate of 63 MHz; therefore, the formation of the black structures was possibly attributed to heat accumulation by the laser pulse train. The peak intensity in the present study is comparable to the peak intensity that induced a change of the refractive index of PDMS under the conditions of a laser wavelength of 800 nm, pulse duration of 130 fs, and repetition rate of 1 kHz, reported in a previous study [15]. Therefore, the femtosecond laser pulse irradiation may induce the scission of chemical bonds.

Figure 1. (**a**) Photographic image of the structure fabricated on the surface of native PDMS. Multiple lines were fabricated with a line–to–line interval of 20 μm by moving the sample at a scanning speed of 2 mm/s in the *x*-direction. The laser power was 150 mW. The size of the scanned area was 5 mm × 5 mm. (**b**) SEM image of the irradiated area on the PDMS surface. The white double-headed arrow shows the scanning direction.

For the measurement of the electrical conductivity of black structures fabricated by irradiation with femtosecond laser pulses, structures with lengths of 8 mm in the *x*-direction and 2 mm in the *y*-direction were fabricated. The structures were fabricated at a scanning speed of 2 mm/s in the *x*-direction. Multiple lines with a line–to–line interval of 25 μm were fabricated, which was sufficient to obtain fully-overlapped black structures. Figure 2 shows the current–voltage curve of the structures. Probes were set 6 mm apart from each other on the surface of the black structures. The current increased linearly with the applied voltage. The average resistance was calculated to be 4.8 kΩ. By assuming that the line structures have a half-circle shape in the cross-section and that they are partially overlapped, the volume of the structure and resistivity are estimated to be 4.0×10^{-5} cm^3 and 5.3 Ω cm, respectively. These results clearly demonstrate that PDMS was modified to electrically conductive material by laser irradiation.

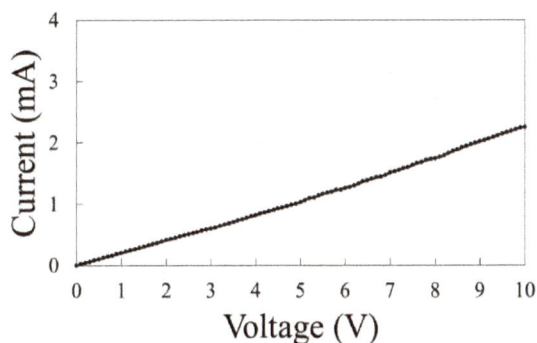

Figure 2. Current–voltage curve of the structures fabricated by laser pulse irradiation. Multiple lines were formed with a line–to–line interval of 25 μm by moving the sample at a scanning speed of 2 mm/s in the *x*-direction. The size of the scanned area was 8 mm in the *x*-direction and 2 mm in the *y*-direction. The laser power was 150 mW.

3.2. Analytical Results of FTIR Spectroscopy and XRD

To investigate the chemical composition of the black structures, FTIR spectroscopy was carried out. Figure 3 shows the FTIR spectra of native PDMS (Figure 3a) and PDMS irradiated by femtosecond laser pulses under the same laser conditions as those in Figure 1 (Figure 3b). In the spectrum of the native PDMS, sharp peaks corresponding to C–H (2950 and 2900 cm^{-1}), CH_2 deformation (1400 cm^{-1}), Si–O (1080 cm^{-1}), Si–CH$_3$ rocking (820 cm^{-1}), and Si–O–Si deformation (460 cm^{-1}) were observed. In the spectrum of the PDMS irradiated by laser pulses, wide peaks of Si–O (1080 cm^{-1}), Si–CH$_3$ rocking (820 cm^{-1}), and Si–O–Si deformation (460 cm^{-1}) were observed; no sharp peak was observed (Figure 3b). The peaks of C–H (2950 and 2900 cm^{-1}) and CH_2 deformation (1400 cm^{-1}), which are typical bonds between carbon and hydrogen, were not observed. The disappearance of the peaks corresponding to C–H (2950 and 2900 cm^{-1}) and CH_2 deformation (1400 cm^{-1}) suggests that scission of the corresponding chemical bonds was induced by the laser pulse irradiation, leading to the release of carbon and hydrogen in the bonds as gaseous species, including hydrocarbon gas and CO_2 gas [16]. Typical peaks of a carbon material, the D band (1350 cm^{-1}) and G band (1598 cm^{-1}) [17], were not observed, showing that the formation of carbon materials is negligible in this study. On the other hand, a peak corresponding to Si–O (1080 cm^{-1}) was observed after the laser pulse irradiation, which indicates the possible formation of SiO_2 and SiO.

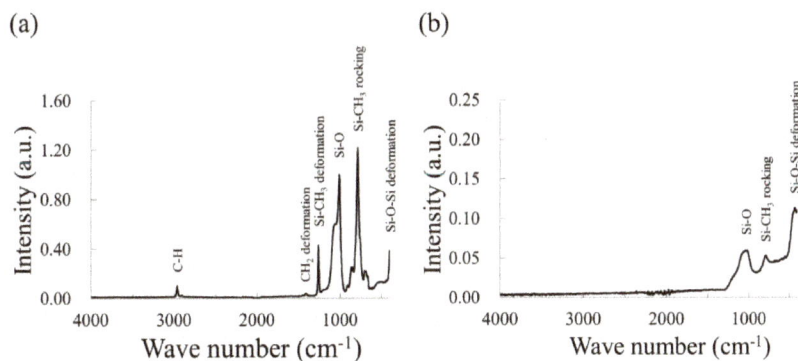

Figure 3. FTIR spectra of (a) native PDMS and (b) PDMS irradiated by femtosecond laser pulses at 150 mW.

In order to identify the material of the black structures, XRD analyses were performed. Figure 4 shows XRD patterns of the native PDMS (Figure 4a) and PDMS irradiated by femtosecond laser pulses (Figure 4b). The laser conditions correspond to the case of Figure 1. For the native PDMS, no significant diffraction peak was observed. On the other hand, diffraction peaks were observed around $2\theta = 36°$, $60°$, and $72°$ for the PDMS irradiated by laser pulses (Figure 4b). These peaks correspond to the (111), (220), and (311) diffraction planes of crystalline β-SiC, demonstrating the laser direct modification from native PDMS to SiC.

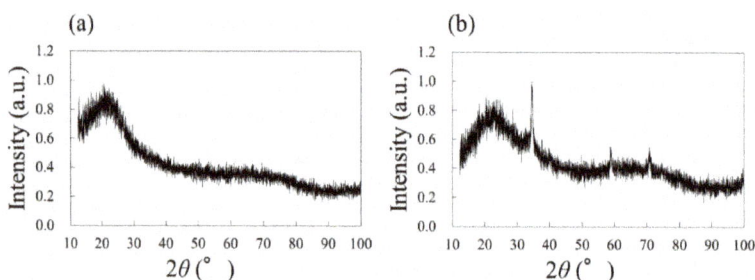

Figure 4. XRD patterns of (**a**) native PDMS and (**b**) PDMS irradiated by femtosecond laser pulses at 150 mW.

The estimated resistivity of the black structures, 5.3 Ω cm (Figure 2), is approximately 40 times larger than the resistivity of bulk SiC, 0.13 Ω cm [18]. The lower electrical conductivity is attributed to the roughness of the formed structures, as well as the possible formation of secondary products, including silicon oxides and silicon carbide oxides. Thermal annealing, including laser-based methods, would improve the electrical conductivity of the formed SiC structures.

Scission of chemical bonds and crystallization are necessary for the formation of β-SiC from PDMS or other siloxanes. Thermal treatments at appropriate temperatures such as oven annealing are reported to be necessary to form SiC [19,20]. It has been widely recognized that consecutive femtosecond laser pulses with a high repetition rate induce heat accumulation [21]. In this study, we used a femtosecond laser pulse train at a repetition rate of 63 MHz, which is considered to be effective in increasing the temperature by heat accumulation on the surface of native PDMS. The formation of SiC from native PDMS by laser irradiation has not been reported previously; however, the formation of β-SiC by laser irradiation of polycarbosilane (PCS) by the sintering of PCS powders using a CO_2 laser [22] or by the pyrolysis of PCS thin films on Si and SiO_2 substrates using a millisecond pulsed laser [23] has been reported. The formation of carbon materials by the irradiation of PDMS (pyrolysis of PDMS using an infrared continuous wave (CW) laser [17]) has also been reported by other groups. In addition to the thermal effect, the photochemical effect induced by the non-linear optical interaction with the high-intensity femtosecond laser pulse should be considered when discussing the scission of chemical bonds and crystallization in this study. For example, the wavelengths of 800 nm and 400 nm led to different results of bond scission with the femtosecond laser ablation, although the sizes of the laser ablation craters were comparable [24].

3.3. Characterization of Structures Formed under Different Irradiation Conditions

Investigation of parameter dependence on the formation of β-SiC by the irradiation of femtosecond laser pulses to native PDMS is crucial for the elucidation of its formation mechanism, leading to the practical application of the presented technique. Figure 5 shows the XRD patterns of structures fabricated by laser pulse irradiation to native PDMS under different irradiation conditions. Note that the XRD analyses shown in Figure 5 were performed for structures formed on PDMS with a 140-μm air-gap between the surface of PDMS and the surface of the cover glass. This setup was performed in order to clarify that Si, required for the formation of SiC, is derived not from the cover glass, but from the PDMS. Under all irradiation conditions, black structures, similar to the structures formed when the PDMS was directly placed on the cover glass, were formed. In Figure 5a–c, the scanning speed was fixed at 2 mm/s and laser power was varied. Diffraction peaks of β-SiC were weak for 70 mW (Figure 5a), while peaks were clear for 150 mW (Figure 5b) and 250 mW (Figure 5c). When laser power was fixed at 150 mW (Figure 5d–f), diffraction peaks of β-SiC were weak for 0.5 mm/s (Figure 5d), while peaks were clear for 1 mm/s (Figure 5e) and 5 mm/s (Figure 5f). The weak diffraction peaks of β-SiC for a lower scanning speed indicate that the exceeding laser pulses may have induced the modification of formed β-SiC to amorphous SiC or other materials.

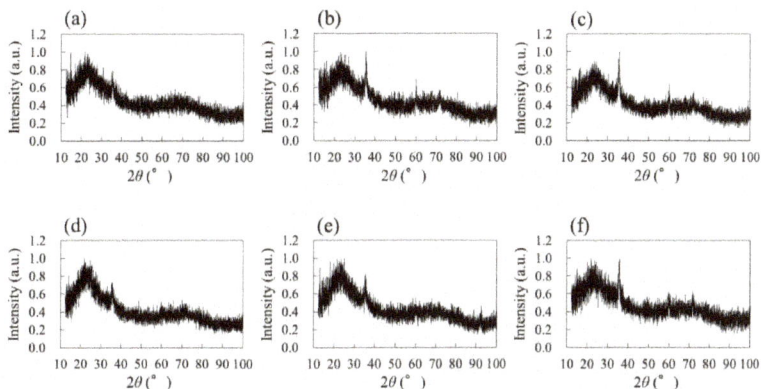

Figure 5. XRD patterns of structures fabricated by laser pulse irradiation to native PDMS under different irradiation conditions. (**a–c**) had a fixed scanning speed of 2 mm/s. Laser power was 70 mW (**a**), 150 mW (**b**), and 250 mW (**c**), respectively. (**d–f**) had a fixed laser power of 150 mW. Scanning speed was 0.5 mm/s (**d**), 1 mm/s (**e**), and 5 mm/s (**f**), respectively.

Figure 6 shows SEM images of structures fabricated by laser pulse irradiation to native PDMS under the same irradiation conditions as Figure 5, respectively. Under the condition of 70 mW at 2 mm/s, where diffraction peaks of β-SiC were weak (Figure 5a), structures aligned along the scanning direction were observed (Figure 6a). Since the laser power was low, it is possible that unmodified PDMS remained and/or that modification might not have been sufficient to form β-SiC. For 150 and 250 mW at 2 mm/s, structures aligned along the scanning direction were also observed, but partially connected (Figure 6b,c). Since the enhanced optical field is generated between adjacent structures on a sub-micro scale [25], localized melting possibly occurred by succeeding laser pulses at the area above the melting temperature of SiC [26] to connect structures. In Figure 6d–f, laser power was fixed at 150 mW, and scanning speed was varied. For 150 mW at 0.5 mm/s, bumps as well as cracks were observed (Figure 6d). The number of pulses per beam spot, i.e., pulse overlap, is calculated to be 2×10^5 pulses at the scanning speed of 0.5 mm/s. For 1 mm/s and 5 mm/s, the surfaces were comparably smooth (Figure 6e,f).

Figure 7 shows the laser power dependence of average resistance for formed structures (Figure 7a) and scanning speed dependence of average resistance for formed structures (Figure 7b). Probes were set 6 mm apart from each other on the surface of fabricated SiC structures. A voltage range of 0 to 10 V was applied and resistance was measured in 0.1 V steps. The average resistance was plotted in the figure. As shown in Figure 7a, average resistance varied with changes in laser power. Among the presented conditions, the lowest resistance was obtained at 150 mW. For higher laser powers, cracks were formed on the surface of SiC structures (Figure 6), which could increase the resistance of the structures. When the laser power was fixed at 150 mW, the lowest resistance was obtained at 2 mm/s. Resistance for scanning speed of 0.5 mm/s was not plotted since hardly any current flowed. With slow scanning speeds, 0.5 mm/s and 1 mm/s, the formation of cracks (Figure 6d,e) and modification of formed β-SiC, possibly due to excessive heat effects, occur, which could decrease the conductivity. For 4 mm/s and 5 mm/s, average resistances of 262.5 and 740.9 kΩ, respectively, were measured. The XRD result shows that the formation of β-SiC is possible even at a scanning speed of 5 mm/s (Figure 5). However, the resistances were higher for the cases of 4 mm/s and 5 mm/s compared to the case of 2 mm/s. Melting of the fabricated structures, which could improve the electrical connection between the formed β-SiC, was assumed to be less than the case for lower scanning speeds, resulting in an increase in resistance. Since the width of the SiC structure depends on the laser power and the scanning speed, further optimization of the scanning interval would be effective in improving the conductivity, as well as forming uniform structures.

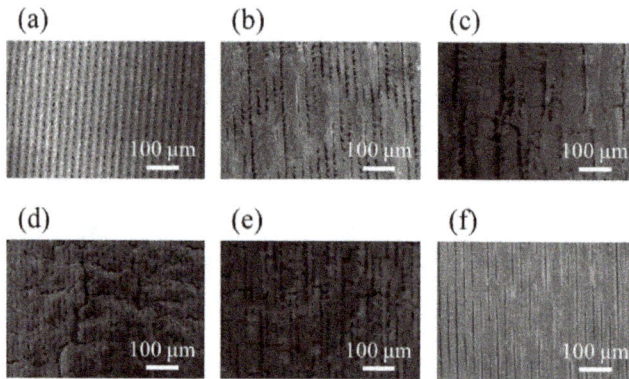

Figure 6. SEM images of SiC surface fabricated on PDMS under different irradiation conditions. (a–c) had a fixed scanning speed of 2 mm/s. Laser power was 70 mW (a), 150 mW (b) and 250 mW (c). (d–f) had a fixed laser power of 150 mW. Scanning speed was 0.5 mm/s (d), 1 mm/s (e) and 5 mm/s (f).

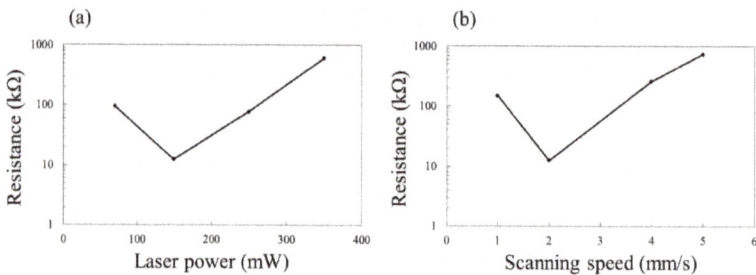

Figure 7. (a) Laser power dependence on average resistance of formed structures. Scanning speed was 2 mm/s. (b) Scanning speed dependence on average resistance of formed structures. Laser power was 150 mW. Hardly any current flowed at 0.5 mm/s.

4. Conclusions

In conclusion, the black structures were formed by the irradiation of femtosecond laser pulses to native PDMS. The formed black structures exhibit electrical conductivity. The FTIR and XRD results showed that the formed black structures were composed of β-SiC. To the best of our knowledge, this is the first demonstration of the direct modification of native PDMS to SiC only by laser irradiation. The technique we present enables simple direct writing of conductive lines on the surface of PDMS, which can be utilized for the fabrication of flexible/stretchable devices.

Author Contributions: Y.N. and S.H. performed experiments. Y.N., S.H., and A.K. prepared the setups for the experiments and measurements. Y.N., N.N., and M.T. designed the study and the project. Y.N. and M.T. wrote the main text of the manuscript. All authors contributed to the discussion and reviewed the manuscript.

Funding: This study was partially supported by a grant from the Amada Foundation. Y.N. is grateful for a Grant-in-Aid for Research Fellow of the Japan Society for the Promotion of Science (JSPS).

Conflicts of Interest: The authors declare no conflict of interest.

References

1. Isiksacan, Z.; Guler, M.T.; Aydogdu, B.; Bilican, I.; Elbuken, C. Rapid fabrication of microfluidic PDMS devices from reusable PDMS molds using laser ablation. *J. Micromech. Microeng.* **2016**, *26*, 035008. [CrossRef]
2. Sekitani, T.; Nakajima, H.; Maeda, H.; Fukushima, T.; Aida, T.; Hata, K.; Someya, T. Stretchable active-matrix organic light-emitting diode display using printable elastic conductors. *Nat. Mater.* **2009**, *8*, 494–499. [CrossRef] [PubMed]
3. Park, H.; Jeong, Y.R.; Yun, J.; Hong, S.Y.; Jin, S.; Lee, S.J.; Zi, G.; Ha, J.S. Stretchable Array of Highly Sensitive Pressure Sensors Consisting of Polyaniline Nanofibers and Au-Coated Polydimethylsiloxane Micropillars. *ACS Nano* **2015**, *9*, 9974–9985. [CrossRef] [PubMed]
4. Martinez, V.; Stauffer, F.; Adagunodo, M.O.; Forro, C.; Vörös, J.; Larmagnac, A. Stretchable Silver Nanowire-Elastomer Composite Microelectrodes with Tailored Electrical Properties. *ACS Appl. Mater. Interfaces* **2015**, *7*, 13467–13475. [CrossRef] [PubMed]
5. Guo, L.; Deweerth, S.P. An effective lift-off method for patterning high-density gold interconnects on an elastomeric substrate. *Small* **2010**, *6*, 2847–2852. [CrossRef] [PubMed]
6. Gou, H.; Xu, J.; Xia, X.; Chen, H. Air Plasma Assisting Microcontact Deprinting and Printing for Gold thin Film and PDMS Patterns. *ACS Appl. Mater. Interfaces* **2010**, *2*, 1324–1330. [CrossRef] [PubMed]
7. He, G.C.; Zheng, M.L.; Dong, X.Z.; Jin, F.; Liu, J.; Duan, X.M.; Zhao, Z.S. The Conductive Silver Nanowires Fabricated by Two-beam Laser Direct Writing on the Flexible Sheet. *Sci. Rep.* **2017**, *7*, 41757. [CrossRef] [PubMed]
8. Terakawa, M.; Torres-Mapa, M.L.; Takami, A.; Hienemann, D.; Nedyalkov, N.N.; Nakajima, Y.; Hordt, A.; Ripken, T.; Hiesterkamp, A. Femtosecond laser direct writing of metal microstructure in a stretchable poly(ethylene glycol) diacrylate (PEGDA) hydrogel. *Opt. Lett.* **2016**, *41*, 1392–1395. [CrossRef] [PubMed]
9. Liu, Y.; Lee, M. Laser Direct Synthesis and Patterning of Silver Nano/Microstructures on a Polymer Substrate. *ACS Appl. Mater. Interfaces* **2014**, *6*, 14576–14582. [CrossRef] [PubMed]
10. Rahimi, R.; Ochoa, M.; Yu, W.; Ziaie, B. Highly stretchable and sensitive unidirectional strain sensor via laser carbonization. *ACS Appl. Mater. Interfaces* **2015**, *7*, 4463–4470. [CrossRef] [PubMed]
11. Rahimi, R.; Ochoa, M.; Ziaie, B. Direct Laser Writing of Porous-Carbon/Silver Nanocomposite for Flexible Electronics. *ACS Appl. Mater. Interfaces* **2016**, *8*, 16907–16913. [CrossRef] [PubMed]
12. Atanasov, P.A.; Stankova, N.E.; Nedyalkov, N.N.; Fukata, N.; Hirsch, D.; Rauschenbach, B.; Amoruso, S.; Wang, X.; Kolev, K.N.; Valova, E.I.; et al. Fs-laser processing of medical grade polydimethylsiloxane (PDMS). *Appl. Surf. Sci.* **2016**, *374*, 229–234. [CrossRef]
13. Stankova, N.E.; Atanasov, P.A.; Nikov, R.G.; Nikov, R.G.; Nedyalkov, N.N.; Stoyanchov, T.R.; Fukata, N.; Kolev, K.N.; Valova, E.I.; Georgieva, J.S.; et al. Optical properties of polydimethylsiloxane (PDMS) during nanosecond laser processing. *Appl. Surf. Sci.* **2016**, *374*, 96–103. [CrossRef]
14. Stankova, N.E.; Atanasov, P.A.; Nedyalkov, N.N.; Stoyanchov, T.R.; Kolev, K.N.; Valova, E.I.; Georgieva, J.S.; Armyanov, S.A.; Amoruso, S.; Wang, X.; et al. Fs- and ns-laser processing of polydimethylsiloxane (PDMS) elastomer: Comparative study. *Appl. Surf. Sci.* **2015**, *336*, 321–328. [CrossRef]
15. Cho, S.H.; Chang, W.S.; Kim, K.R.; Hong, J.W. Femtosecond laser embedded grating micromachining of flexible PDMS plates. *Opt. Commun.* **2009**, *282*, 1317–1321. [CrossRef]
16. Hasegawa, Y.; Iimura, M.; Yajima, S. Synthesis of continuous silicon carbide fibre—Part 2 Conversion of polycarbosilane fibre into silicon carbide fibres. *J. Mater. Sci.* **1980**, *15*, 720–728. [CrossRef]
17. Alcántara, J.C.C.; Zorrilla, M.C.; Cabriales, L.; Rossano, L.M.L.; Hautefeuille, M. Low-cost formation of bulk and localized polymer-derived carbon nanodomains from polydimethylsiloxane. *Beilstein J. Nanotechnol.* **2015**, *6*, 744–748. [CrossRef] [PubMed]

18. Samsonov, G.V. *Plenum Press Handbooks of High Temperature Materials No. 2 Properties Index*; Plenum Press: New York, NY, USA, 1965; pp. 1–635, ISBN 978-1-4899-6405-2.
19. Al-Ajrah, S.; Lafdi, K.; Liu, Y.; Le Coustumer, P. Fabrication of ceramic nanofibers using polydimethylsiloxane and polyacrylonitrile polymer blends. *J. Appl. Polym. Sci.* **2018**, *135*, 45967. [CrossRef]
20. Burns, G.T.; Taylor, R.B.; Xu, Y.; Zangvil, A.; Zank, G.A. High-Temperature Chemistry of the Conversion of Siloxanes to Silicon Carbide. *Chem. Mater.* **1992**, *4*, 1313–1323. [CrossRef]
21. Eaton, S.; Zhang, H.; Herman, P.; Yoshino, F.; Shah, L.; Bovatsek, J.; Arai, A. Heat accumulation effects in femtosecond laser-written waveguides with variable repetition rate. *Opt. Express* **2005**, *13*, 4708–4716. [CrossRef] [PubMed]
22. Jakubenas, K.; Marcus, H.L. Silicon Carbide from laser Pyrolysis of Polycarbosilane. *J. Am. Chem. Soc.* **1995**, 2263–2266. [CrossRef]
23. Colombo, P.; Martucci, A.; Fogato, O.; Villoresi, P. Silicon Carbide Films by Laser Pyrolysis of Polycarbosilane. *J. Am. Ceram. Soc.* **2001**, *26*, 224–226. [CrossRef]
24. Shibata, A.; Machida, M.; Kondo, N.; Terakawa, M. Biodegradability of poly(lactic-co-glycolic acid) and poly(l-lactic acid) after deep-ultraviolet femtosecond and nanosecond laser irradiation. *Appl. Phys. A* **2017**, *123*, 438. [CrossRef]
25. Terakawa, M.; Nedyalkov, N.N. Near-field optics for nanoprocessing. *Adv. Opt. Technol.* **2016**, *5*, 17–28. [CrossRef]
26. Ashwath, P.; Xavior, M.A. Processing methods and property evaluation of Al_2O_3 and SiC reinforced metal matrix composites based on aluminium 2xxx alloys. *J. Mater. Res.* **2016**, *31*, 1201–1219. [CrossRef]

nanomaterials

MDPI

Article

Extreme Energy Density Confined Inside a Transparent Crystal: Status and Perspectives of Solid-Plasma-Solid Transformations

Eugene G. Gamaly [1],*, Saulius Juodkazis [2] and Andrei V. Rode [1],*

[1] Laser Physics Centre, Research School of Physics and Engineering, The Australian National University, Canberra ACT 2601, Australia

[2] Centre for Micro-Photonics, Swinburne University of Technology, Hawthorn VIC 3122, Australia; sjuodkazis@swin.edu.au

* Correspondence: eugene.gamaly@anu.edu.au (E.G.G.); andrei.rode@anu.edu.au (A.V.R.); Tel.: +61-2-6125-8659 (E.G.G.); +61-2-6125-4637 (A.V.R.)

Received: 26 June 2018; Accepted: 19 July 2018; Published: 21 July 2018

Abstract: It was demonstrated during the past decade that an ultra-short intense laser pulse tightly-focused deep inside a transparent dielectric generates an energy density in excess of several MJ/cm^3. Such an energy concentration with extremely high heating and fast quenching rates leads to unusual solid-plasma-solid transformation paths, overcoming kinetic barriers to the formation of previously unknown high-pressure material phases, which are preserved in the surrounding pristine crystal. These results were obtained with a pulse of a Gaussian shape in space and in time. Recently, it has been shown that the Bessel-shaped pulse could transform a much larger amount of material and allegedly create even higher energy density than what was achieved with the Gaussian beam (GB) pulses. Here, we present a succinct review of previous results and discuss the possible routes for achieving higher energy density employing the Bessel beam (BB) pulses and take advantage of their unique properties.

Keywords: light-matter interaction; ultra-short laser pulses; high-pressure/density conditions; phase transitions

---

1. Microexplosion Studies with Gauss-Shaped Beam

The studies of confined microexplosions during the last decade revealed the major features of this complicated phenomenon where the processes of electro-magnetic field/dielectric interaction, plasma formation and high-pressure hydrodynamics are intertwined. The concise description of these processes is as follows. The tight focusing of the laser beam deep inside a transparent crystal allows achieving the absorbed energy density in excess of the strength of any material in a sub-micron volume surrounded by the pristine solid. After energy transfer from hot electrons to ions, the expanding strong shock wave accompanied by the rarefaction wave starts propagating outside of this volume. After the shock decelerating and stopping, the void, surrounded by a shell of compressed and pressure modified material converted to the novel phases, is formed. All transformed material remains confined inside the bulk of undamaged material ready for further studies. These studies employed the short intense laser beam with the Gaussian spatial and temporal intensity profile [1–4].

The short intense laser pulse with the Gaussian spatial and temporal intensity profiles tightly focussed inside a transparent crystal generates an energy density of several MJ/cm^3. The pressure produced is in excess of a few TPa, which is higher than the strength of any existing material (diamond has the highest Young's modulus of 1 TPa = 1 MJ/cm^3). The laser pulse, 150 fs, 100–200 nJ, 800 nm, tightly-focussed inside sapphire with a microscope lens ($NA = 1.4$) creates the solid density plasma at the temperature of a few tens of electron Volts ($\sim 5 \times 10^5$ K) with the record-high heating rate of 10^{18} K/s [1,2]. It was found that the novel (previously unobserved) high-pressure phases of aluminium and silicon were formed [3,4] following the ultrashort laser-induced confined microexplosion. Pressure/temperature conditions created in the microexplosion are similar to those in hot cores of stars and planets ("primeval soup" or warm dense matter). The material converted to high pressure/temperature solid density plasma is then transformed into the novel solid phase during the ultra-fast cooling and re-structuring. The major difference from the core-star conditions is the record-fast cooling ($\sim 10^{16}$ K/s) from plasma state to solid state. In the previous experiments, the study of the pressure-affected materials was produced postmortem, well after the end of the pulse when transformed material was cooled down to the ambient conditions. The structure of laser-transformed material was determined by the synchrotron X-ray diffraction [3] and with the electron diffraction [4] in transmission electron microscopy (TEM) [4].

1.1. Novelty of the Phase Transformation Path during and after Confined Microexplosion

The solid transforms to a solid-density plasma state ($T_e \sim 50$ eV) during a pulse time shorter than all energy relaxation times. A strong shock wave (SW) starts propagating from the energy deposition region several picoseconds after the pulse due to energy transfer from electrons to the ions. The shock wave decelerates and converts into a sound wave in the surrounding cold pristine crystal. The phenomenon is similar, but not identical to an underground nuclear explosion: the massless energy carriers (photons) deliver the energy inside a transparent crystal without changing the atomic and mass content of a material. All laser-affected material is expelled from the energy deposition area by the combined action of shock and rarefaction waves, forming a void surrounded by the shell of material compressed against the surrounding cold pristine crystal. The material returns from the high-pressure plasma state (high entropy, chaotic) to the ambient conditions at room temperature/pressure, however attaining a phase state different from the initial solid state. In all known methods of high pressure phase formation, the initial crystalline structure is re-structured, i.e., the atoms are moved from the initial arrangement to the new positions under the action of high pressure. During the transformation path under confined microexplosion, the initial state of a crystal is completely destroyed and forgotten. The irradiated material is converted into a chaotic mixture of ions and electrons at high temperature. Therefore relaxation to the ambient conditions occurs along the unknown paths going through the metastable intermediate equilibrium potential minima. The theoretical (computational, modified DFT-studies) during the last decade searched for the possible paths of material transformations under high pressure from the initially chaotic (stochastic) state [5]. These studies uncovered many physically allowed paths for the formation of multiple novel phases (including incommensurable phases) from the initially chaotic state. The confined microexplosion method now is the only practically realised way for the formation of novel material phases from the plasma state, preserving the transformed material confined inside the pristine crystal for further structural studies.

1.2. Limitations of the Confined Microexplosion Method with the GB

There are limitations to the energy density and amount of laser-affected material in confined microexplosion generated by the tightly-focused Gauss beam. The main limitation is imposed by diffraction: the radius of the diffraction-limited focal spot is [2,6]: $r_{Airy} = 0.61\lambda/NA$; which defines the central Airy disk at the focus. For $\lambda = 800$ nm and $NA = 1.4$, one gets $r_{foc} = 0.35$ μm and a focal area of 0.38 μm^2. The absorption length in dense plasma equals ~ 30 nm, giving the energy deposition volume $\sim 10^{-14}$ cm^{-3}. With absorbed energy around 100 nJ, the absorbed energy density amounts to 10^7 J/cm^3 = 10 TPa. The number of laser-affected atoms constitutes around 10^{11} atoms (a few picograms), making structural studies extremely difficult. Therefore, the questions arises: is it possible to increase the absorbed energy density and/or increase the amount of the laser-affected material and thus the amount of the novel phase? Preliminary studies have shown that it is very difficult to overcome the energy density of several MJ/cm^3 (several TPa of pressure) and increase the amount of laser-affected material using a tightly-focused Gauss beam. First, the ionisation wave moving towards the laser pulse with increasing intensity increases the absorbing volume and limits the energy density [7]. Moreover, the experiments with increasing laser pulse energy demonstrated that at the energy per 150-fs pulse of 200 nJ, the cracks surrounding the focal area destroyed the regular void formation [2]. Diffraction-free Bessel beams (BB) raised the hope of achieving a higher energy density and larger amounts of the material affected. Below, we describe the recent progress made with these studies. Then, we describe some effects (and unresolved problems), the solutions of which may lead to a further increase of the absorbed energy density.

2. Status of the BB-Transparent Crystal Interactions

It was demonstrated recently that the BB (150 fs, 2 µJ) focused inside sapphire produced a cylindrical void of 30 µm in length and 300 nm in diameter [8]. The void volume, $V = 30$ µm $\times \pi r^2 = 2.12 \times 10^{-12}$ cm^{-3}, appears to be two orders of magnitude larger than that generated by the GB. The conclusions based solely on the void size measurements and on the energy and mass conservation laws without any ad hoc assumptions about the interaction process are the following [9]. The material initially filling the void was expelled and compressed into a shell by the high-pressure shock wave. The work necessary to remove material with the Young modulus Y from volume V equals at least $Y \times V = 0.848$ µJ ($Y = 4 \times 10^5$ J/cm^3, the Young modulus of sapphire). This is evidence of strong (>40%) absorption of the pulse energy. In order to generate a strong shock wave capable of expelling such an amount of material, the absorbed energy should be concentrated in the central spike with a much smaller diameter than that of the void (the absorbed energy density is still not known, theoretically nor experimentally).

The unique features of the diffraction-free Bessel beam spatial distribution of intensity in the focal area allow one to understand some of the experimental findings and indicate new problems and opportunities. The spatial distribution of intensity across the cylindrical focal volume in a transparent medium unaffected by light and observed experimentally is close to Durnin's solution, $J_0^2(k_r r)$ [10]: the central spike surrounded by circular bands with the maximum of intensity on the axis approximately five-times higher than in the next band.

The parameters of the quasi non-diffracting BB created by any device from the incoming cw-laser pulse in air (axicon, circular slit, spatial light modulator (SLM), etc.) with the cone angle θ are the following (Figure 1): radius of the incoming beam before the BB-creating device, R; $z_{max} = R/\tan(\theta)$; $k_r = k \times \sin(\theta)$; $k_z = k \times \cos(\theta)$ [10]. Building the intensity distribution in the low intensity short pulse BB occurs in a similar way to that as with the cw-laser, as was demonstrated experimentally [8]. It is worth noting that in these experiments, during the pulse time, the beam propagates a distance comparable to the length of the elongated cylindrical focus (Z_{max}), $L_{pulse} = t_{pulse} \times c/n$ (n is the refractive index in a transparent crystal unaffected by a laser). For example, a 150-fs (800 nm) low intensity pulse in sapphire propagates ~30 microns, which is close to $t_{pulse} \times c/n = 30$ µm [8].

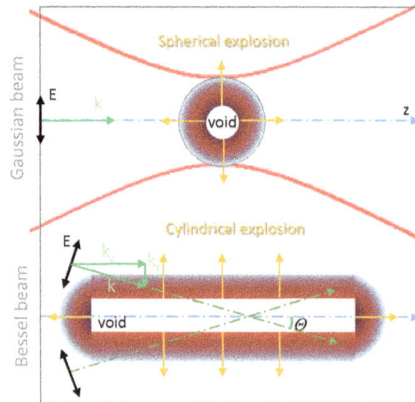

Figure 1. Schematic presentation of radial microexplosion-driven inside a transparent material by a focused linearly-polarised (E-field) Gaussian beam (GB) and Bessel beam (BB) with projection E_z along the optical axis (z-axis); θ is the angle with the optical axis (wavevector $k = \sqrt{k_r^2 + k_z^2}$ is shown on the upper half of the conical wave). Resonant absorption of the E_r component (along the radial direction k_r) allows a higher energy density deposition for the cylindrical microexplosion. With the central void diameter in sapphire comparable for the GB [1] and BB [8] pulses, the volume was 180-times larger in the case of BB.

In the non-absorbing media, the length of the focus (the distance where diffraction is strongly suppressed) apparently is independent of the pulse duration. On the axis of the BB (in the focal area), the time of the interaction of the electromagnetic wave with matter could be shorter than the pulse duration. Therefore, a beam of any duration allegedly propagates the same distance Z_{max} allowed by the focusing device. This seemingly obvious statement should be confirmed experimentally.

Under the action of intense pulse, the ionization breakdown occurs early in the pulse time near the central spike where intensity is maximum. The studies of the interaction process of intense BB at an intensity above the ionization threshold are absent to the best of our knowledge. Estimates, suggestions and problems relating to the formation of the intensity distribution and interaction process based on the studies of confined microexplosion and intense short pulse interactions with dielectrics are presented below. Experimental observation of the void formation by GB and BB pulses is shown in Figure 2. The BB pulses are used to dice transparent materials [11] and to inscribe high efficiency optical gratings in silica [12].

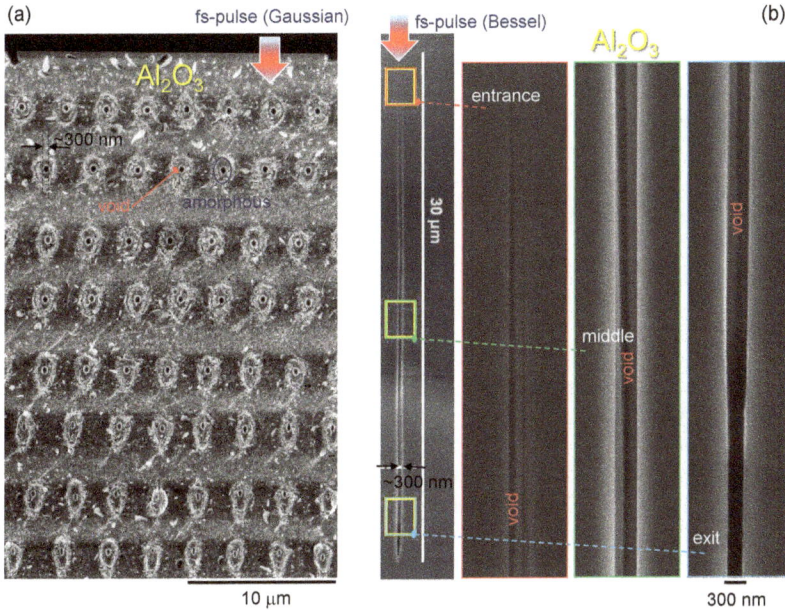

Figure 2. SEM side-view images of the voids made with ultra-short Gaussian [1,13] (a) and Bessel [8] (b) single pulses in sapphire. Focusing of the 800-nm/130-fs Gaussian pulses of ~150 nJ of energy was carried out with an objective lens of numerical aperture $NA = 1.3$ and was stacked into a vertical plane of the void-structures [13]. This plane was used to split the sapphire sample for the side-view SEM observation. The voids made at larger depth were affected by spherical aberration, which reduced the void and elongated amorphous region. The 800-nm/140-fs, 2 µJ of energy, Bessel pulses were used to make cylindrical voids of a diameter of ~ 300 nm revealed by focussed ion beam (FIB) milling [8].

Control of Energy Deposition by BB Pulses

In short intense BB interaction with the initially transparent medium, the ionization threshold is reached at the axis of the focal volume where intensity attains the maximum value. It occurs early in the pulse time close to the beginning of the elongated focal region. A narrow cylindrical plasma region is created along the axis. Incident light starts absorbing in plasma. Let us take the incident field structure near the axis as the following E (E_r, $E_\varphi = 0$; E_z); H (0; H_φ, 0). Then, the Poynting vector reads, $\mathbf{S} = \frac{c}{4\pi}(\mathbf{E} \times \mathbf{H})$. Therefore, the energy flows are generated inward along the radius and along the z-axis in direction of the beam propagation: $S_r = \frac{c}{4\pi}(H_\varphi \cdot E_z)$ and $S_z = \frac{c}{4\pi}(H_\varphi \cdot E_r)$ [J/(cm^{2}s)].

Thus, by changing the cone angle, one can control the radial and axial energy flows. The interaction mode of intense BB with a transparent crystal dramatically changes after the ionization threshold is achieved. The surface, where the real part of the permittivity is zero, $\varepsilon_{re} = 0$, separates the dielectric ($\varepsilon_{re} > 0$) and plasma ($\varepsilon_{re} < 0$) regions. The gradient of the permittivity is directed along the radius of a cylinder. The energy flow goes inward in the radial direction. Thus, the incident wave splits into the evanescent and reflection waves. The resonance absorption occurs in the vicinity of the zero-epsilon surface, creating a plasma wave (plasmon) propagating along the radius in the direction to the axis of the cylindrical focal region. The evanescent wave decays along the radius in the same direction. Thus, the zero-permittivity surface generates simultaneously coherent plasmons and evanescent waves coming together (focusing) to the axis of the cylindrical focal region. One may expect that coupling of evanescent waves and plasmons also contributes to the increase of the intensity and energy density near the axis in a way similar to that discovered in the studies of extraordinary optical transmission (EOT) through sub-wavelength hole arrays [14].

The ideal diffraction free beam is the monochromatic Bessel beam [10], created via superposition of plane waves, the wave vectors of which are evenly distributed over the surface of a cone. The Bessel function of the first kind zero order, J_0, is a sum of the Hankel functions of the first and second kind [15], where the inward energy flow is balanced by the outward flow.

It was suggested [16] that the quasi diffractionless BB can be presented as the result of the interference of two conical running Hankel beams, carrying equal amounts of energy towards and outwards from the beam axis, yielding no net transversal energy flux in the BB. The interference of two Hankel beams with different amplitudes creates unbalanced BB where the net radial energy flux appears. Unbalancing creates the inward radial energy flux from the conical tails of the beam. The study of stability in the frame of the non-linear Shrodinger equation (NLSE) equation revealed that the Bessel-like solutions in pure Kerr media are unstable [17].

In the interaction of intense short pulse BB with a transparent dielectric at the intensity below the ionization threshold, the BB apparently retains its balanced structure. After the plasma formation, the energy flow directed inward to the axis is created due to absorption leading to destruction of this balance. One may argue that after the ionization threshold, the Hankel function of the first kind might be considered as an appropriate approximation of the field distribution near the axis of cylindrical focus, being the exact solution of the Bessel equation describing the electric field increasing while focusing. One may conjecture that the BB becomes unstable, tending to focus onto the cylindrical axis, thus creating an energy density higher than a tightly-focused, but diffraction-limited Gaussian beam.

Let us now consider the relation between the pulse duration, absorption, focal region length and laser-affected area length. In short intense BB interaction with the initially transparent medium, the ionization threshold is reached at the axis of the focal volume where the intensity is maximum. This occurs early in the pulse time close to the beginning of the elongated focal region. The intense pulse converts the initially transparent material into strongly absorbing plasma practically at the moment of its arrival at some space point. Therefore, the plasma region gradually increases along the axis as the pulse proceeds until the end of the pulse. The last portion of light arrives after travelling through the transparent crystal a distance $t_p \times c/n$. The laser-affected distance then reads $L_{las} = (t_p \times c/n) \cos \theta$ (θ is the half-cone angle; Figure 1). One can see now the difference between the BB-affected area in a transparent medium (diffraction-free focus) and the laser-affected area in an intense short pulse laser/crystal interaction. For sufficiently short pulses, the laser-affected area might be shorter than the diffraction-free zone, $L_{las} < Z_{max}$. Thus, laser pulse duration might be another lever (along with the cone angle) to control the energy deposition volume.

Experiments demonstrated that short intense BB could affect a much larger amount of material producing solid-plasma-solid transformation (direct measurements) at allegedly a pressure of several TPa (conclusions on the basis of the analysis of the experiments) [8,9]. J.Hu [18] measured the average speed of the shock wave, $v_{sw} \approx 60$ km/s, during the cylindrical microexplosion, generated by the BB in sapphire, by the pump-probe technique. The estimate of the driving pressure based on this measurement, $P_{sw} = \varrho_0 v_{sw}^2 = 14.4$ TPa (ϱ_0 is the initial mass density of sapphire), gives the direct experimental evidence of the extreme energy density created by the BB in the focal volume.

There are indications from theoretical studies [9] that the originally stable diffraction-free BB at high intensity in the presence of strong ionization nonlinearity may become unstable. Now, it is difficult to conclude if this may happen in a way similar to the self-focusing instability with Kerr-like non-linearity (rather, not because the paraxial approximation is invalid in this case) or similar to the instability of two unbalanced Hankel beams, which seems more relevant to the case (again, the ionization non-linearity should be accounted for).

The oblique incidence, inherent for the formation of the BB and long focus, implies the possibility of the surface wave (plasmon) formation and propagation along the zero-real-permittivity surface at the same time with the plasmon moving radially due to the resonance absorption. The plasma wave may converge to the axis, contributing to the increase in the absorbed energy density. One may conjecture if it might be relevant for some kind of Langmuir collapse.

It would be crucially important to find the electric field distribution up to the central axis in order to determine the absorbed energy density. This requires a solution of the Maxwell equations in cylindrical geometry coupled to material equations accounting for the change in the permittivity (electrons" number density and collision rate) in accord with the intensity in any space/time point. This is a formidable task; however, it can be clearly formulated for the numerical solution. Different approximations may also be discussed.

It was demonstrated experimentally [18] that the Bessel beam-induced microexplosion in sapphire, producing open-ended channel, proceeds as an axial-symmetric cylindrical explosion, and a mass conservation was experimentally validated [19]. Therefore, the direct theoretical modeling of the cylindrical explosion after the energy deposition of the BB beam inside a narrow on-axis cylinder also can be performed in the frame of two-temperature plasma hydrodynamics in cylindrical geometry in a way similar to as was done with the Gauss beam in spherical geometry [2].

3. Conclusions and Outlook

In conclusion, we should state that further progress in achieving and steering the high energy density strongly depends on the future pump-probe experiments, which will register with time/space resolution the history of the BB-generated microexplosion, processes of returning to the ambient state and new phases' formation. It is worth showing the time and space scales for the succession of events comprising such a history that might in some approximation be extracted from the previous studies [2,4,9].

Let us suggest the BB, 2 μJ, 800 nm, 150 fs, impinges a sapphire crystal several tens of microns thick, creating a focal region of ~30 μm long at the ten microns depth from the outer surface of a sample. The stages of successive transformations are the following; the time count starts at the beginning of the pump pulse:

1. The low intensity stage before ionisation threshold lasts a few fs at the beginning of the pulse;
2. As the ionisation threshold is attained, the cylindrical plasma region is created at the axis of the focal region with a diameter less than a micron. One should note that the full length of the focal region of 30 μm is reached at the end of the pulse, assuming that light propagates as in unaffected sapphire with a speed of $c/n \sim 2 \times 10^{10}$ cm/s;
3. The cylinder diameter of the energy absorption region to the end of the pulse allegedly might be around the doubled absorption length in a dense plasma ~60 nm;
4. The shock wave is created after the energy transfer from electrons to ions in a 7–10-ps time span;
5. The shock wave propagates during another 4–6 ps until it is converted into the acoustic wave, effectively stopped by the cold pressure of the crystal (~Young modulus of sapphire). The void surrounded by the shell of compressed material is formed by the rarefaction wave;
6. The thermal wave of conventional heat conduction spreads into the laser-unaffected crystal, cooling the laser-affected area down to the ambient conditions during tens of nanoseconds. The material re-structuring occurs most probably during Stages 5 and 6. The whole area affected by the heat from the laser-heated region is a cylinder with a length of around 32–34 microns with a diameter of about 2–4 micrometers.

Thus, the whole area affected by the shock and heat waves from the energy deposition region is a cylinder 30 microns long and a few microns in diameter. The time span for the whole process of material transformation is around tens of nanoseconds. The recent arrival of X-ray free electron lasers (XFEL) with a pulse duration as short as 7–15 fs and a photon energy of 8–10 keV currently available at EuroXFEL at DESY in Hamburg and in SACLA XFEL at Spring-8 at Riken Institute in Japan creates new opportunities for uncovering the mechanism of the formation of the new states of matter. Up to 17-keV pulses expected in the near future at the SLAC National Acceleration Laboratory at Stanford.

All these new sources or coherent ultra-short X-ray radiation will be used to uncover the processes involved in formation of such unusual material states. For such experiments, the tailored axial intensity distribution of the optical BB pulses can be prepared using diffraction optical elements [20], which can be made with a central hole for the co-axial fs-optical-pump and fs-X-ray-probe. To conclude, a light (or X-ray) probe with a sub-picosecond duration and sub-micron spatial resolution may shed light on the unusual formation of novel high-pressure phases starting from the "primeval soup" (warm dense matter) to the solid state at the ambient conditions, being preserved and confined inside a bulk of pristine crystal ready for further structural studies [7].

Author Contributions: E.G.G. conceived the idea, A.V.R. and S.J. carried out experiments. E.G.G. wrote the first draft. All authors edited the final version.

Funding: The Australian Research Council Discovery project DP170100131.

Acknowledgments: This research was supported by the Australian Government through the Australian Research Council's Discovery Projects scheme.

Conflicts of Interest: The authors declare no conflict of interest.

References

1. Juodkazis, S.; Nishimura, K.; Tanaka, S.; Misawa, H.; Gamaly, E.E.; Luther-Davies, B.; Hallo, L.; Nicolai, P.; Tikhonchuk, V. Laser-induced microexplosion confined in the bulk of a sapphire crystal: Evidence of multimegabar pressures. *Phys. Rev. Lett.* **2006**, *96*, 166101. [CrossRef] [PubMed]
2. Gamaly, E.E.; Juodkazis, S.; Nishimura, K.; Misawa, H.; Luther-Davies, B.; Hallo, L.; Nicolai, P.; Tikhonchuk, V. Laser-matter interaction in a bulk of a transparent solid: confined microexplosion and void formation. *Phys. Rev. B* **2006**, *73*, 214101. [CrossRef]
3. Vailionis, A.; Gamaly, E.G.; Mizeikis, V.; Yang, W.; Rode, A.; Juodkazis, S. Evidence of super-dense Aluminum synthesized by ultra-fast microexplosion. *Nat. Commun.* **2011**, *2*, 445. [CrossRef] [PubMed]
4. Rapp, L.; Haberl, B.; Pickard, C.J.; Bradby, J.E.; Gamaly, E.G.; Williams, J.S.; Rode, A.V. Experimental evidence of new tetragonal polymorphs of silicon formed through ultrafast laser-induced confined microexplosion. *Nat. Commun.* **2015**, *6*, 7555. [CrossRef] [PubMed]
5. Shi, X.; He, C.; Pickard, C.J.; Tang, C.; Zhong, J. Stochastic generation of complex crystal structures combining group and graph theory with application to carbon. *Phys. Rev. B* **2018**, *97*, 014104. [CrossRef]
6. Born, M.; Wolf, E. *Principles of Optics*; Cambridge University Press: Cambridge, UK, 2003.
7. Gamaly, E.G.; Rapp, L.; Roppo, V.; Juodkazis, S.; Rode, A.V. Generation of high energy density by fs-laser induced confined microexplosion. *New J. Phys.* **2013**, *15*, 025018. [CrossRef]
8. Rapp, L.; Meyer, R.; Giust, R.; Furfaro, L.; Jacquot, M.; Lacourt, P.A.; Dudley, J.M.; Courvoisier, F. High aspect ratio microexplosions in the bulk of sapphire generated by femtosecond Bessel beams. *Sci. Rep.* **2016**, *6*, 34286. [CrossRef] [PubMed]
9. Gamaly, E.G.; Rode, A.V.; Rapp, L.; Giust, R.; Furfaro, L.; Lacourt, P.A.; Dudley, J.M.; Courvoisier, F.; Juodkazis, S. Interaction of the ultra-short Bessel beam with transparent dielectrics: Evidence of high-energy concentration and multi-TPa pressure. *arXiv* **2017**, arXiv:1708.07630.
10. Durnin, J.; Miceli, J.J.; Eberly, J.H. Diffraction-free beams. *Phys. Rev. Lett.* **1987**, *58*, 1499. [CrossRef] [PubMed]
11. Marcinkevičius, A.; Juodkazis, S.; Matsuo, S.; Mizeikis, V.; Misawa, H. Application of Bessel Beams for microfabrication of dielectrics by femtosecond laser. *Jpn. J. Appl. Phys.* **2001**, *40*, L1197–L1199. [CrossRef]
12. Mikutis, M.; Kudrius, T.; Šlekys, G.; Paipulas, D.; Juodkazis, S. High 90% efficiency Bragg gratings formed in fused silica by femtosecond Gauss-Bessel laser beams. *Opt. Mat. Express* **2013**, *11*, 1862–1871. [CrossRef]
13. Buividas, R.; Gervinskas, G.; Tadich, A.; Cowie, B.; Mizeikis, V.; Vailionis, A.; de Ligny, D.; Gamaly, E.G.; Rode, A.; Juodkazis, S. Phase transformation in laser-induced microexplosion in olivine (Fe,Mg)$_2$SiO$_4$. *Adv. Eng. Mater.* **2014**, *16*, 767–773. [CrossRef]
14. Liu, H.; Lalanne, P. Microscopic theory of the extraordinary optical transmission. *Nature* **2008**, *452*, 728–731. [CrossRef] [PubMed]
15. Morse, P.; Feshbach, H. *Methods of Theoretical Physics*; McGraw Hill: New York, NY, USA, 1953; Volume 1–2.

16. Salo, J.; Fagerholm, J.; Friberg, A.T.; Salomaa, M.M. Unified description of nondiffracting X and Y waves. *Phys. Rev. E* **2000**, *62*, 4261. [CrossRef]

17. Porras, M.A.; Parola, A.; Faccio, D.; Dubietis, A.; Trapani, P.D. Nonlinear Unbalanced Bessel Beams: Stationary Conical Waves Supported by Nonlinear Losses. *Phys. Rev. Lett.* **2004**, *93*, 153902. [CrossRef] [PubMed]

18. Hu, J. High throughput micro/nano manufacturing by femtosecond laser temporal pulse shaping. In Proceedings of the 9th International Conference on Information Optics and Photonics, Harbin, China, 17–20 July 2017.

19. Wang, G.; Yu, Y.; Jiang, L.; Li, X.; Xie, Q.; Lu, Y. Cylindrical shockwave-induced compression mechanism in femtosecond laser Bessel pulse micro-drilling of PMMA. *Appl. Phys. Lett.* **2017**, *110*, 161907. [CrossRef]

20. Dharmavarapu, R.; Bhattacharya, S.; Juodkazis, S. Diffractive optics for axial intensity shaping of Bessel beams. *J. Opt.* **2018**, *20*, 085606. [CrossRef]

nanomaterials

MDPI

Article

Liquid-Assisted Femtosecond Laser Precision-Machining of Silica

Xiao-Wen Cao [1], Qi-Dai Chen [2], Hua Fan [2], Lei Zhang [1], Saulius Juodkazis [3,4] and Hong-Bo Sun [2,5,*]

[1] State Key Laboratory of Integrated Optoelectronics, School of Mechanical Science and Engineering, Nanling Campus, Jilin University, Changchun 130025, China; xw2015@outlook.com (X.-W.C); zhanglei@jlu.edu.cn (L.Z.)
[2] State Key Laboratory of Integrated Optoelectronics, College of Electronic Science and Engineering, Jilin University, Changchun 130012, China; chenqd@jlu.edu.cn (Q.-D.C.); fanhua17@mails.jlu.edu.cn (H.F.)
[3] Centre for Micro-Photonics, Faculty of Science, Engineering and Technology, Swinburne University of Technology, Hawthorn, VIC 3122, Australia; sjuodkazis@swin.edu.au
[4] Melbourne Centre for Nanofabrication, ANFF, 151 Wellington Road, Clayton, VIC 3168, Australia
[5] State Key Laboratory of Precision Measurement Technology and Instruments, Department of Precision Instrument, Tsinghua University, Haidian, Beijing 100084, China
* Correspondence: hbsun@tsinghua.edu.cn; Tel: +86-010-6279-8249

Received: 12 April 2018; Accepted: 23 April 2018; Published: 28 April 2018

Abstract: We report a systematical study on the liquid assisted femtosecond laser machining of quartz plate in water and under different etching solutions. The ablation features in liquid showed a better structuring quality and improved resolution with 1/3~1/2 smaller features as compared with those made in air. It has been demonstrated that laser induced periodic structures are present to a lesser extent when laser processed in water solutions. The redistribution of oxygen revealed a strong surface modification, which is related to the etching selectivity of laser irradiated regions. Laser ablation in KOH and HF solution showed very different morphology, which relates to the evolution of laser induced plasma on the formation of micro/nano-features in liquid. This work extends laser precision fabrication of hard materials. The mechanism of strong absorption in the regions with permittivity (epsilon) near zero is discussed.

Keywords: femtosecond laser; silica; Laser materials processing; nonlinear optics at surfaces

1. Introduction

Femtosecond (fs) laser has proved to be an efficient tool for micro/nanomachining [1] in micro-optics [2], micromechanics [3], microfluidics [4], organic light-emitting diode (OLED) display [5], and micro-sensing [6]. Based on the nonlinear nature of light–matter interaction via multi-photon and avalanche absorption [4,7], fs-laser machining is independent of the material's hardness and has been demonstrated on a wide range of metals, semiconductors, and dielectrics [8–10]. By intense laser pulses focused with an objective lens, structures and patterns of 2D and 3D morphology have been realized [2,8]. However, it also has disadvantages. The laser–matter interaction at the surface is inevitably affected by the evolving ablation pattern of the fabricated structure due to the scattering and absorption, also chemical modification [11]. The laser-induced ripples generated by an imprint of a plasmonic wave [12,13] would greatly increase the surface roughness which can be desired depending on application. Also, laser ablated debris randomly falling on the surface could enhance the light absorption and scattering [9,14]. Especially when laser induced thermal effects are pronounced [15,16], debris is difficult to wash out after laser fabrication. To solve these problems, fs-laser-assisted etching method [17,18] and liquid-assisted fs-laser machining [19,20] have been proposed as two possible solutions. The former method uses laser-induced selective etching to remove the modification area without leaving debris. While the latter uses the laser induced shock wave and water plasma generating cavitation bubbles to wash the debris. When a laser beam is focused on the sample surface through the liquid, subsequent laser pulses are scattered by the bubbles which is a disadvantage for precision machining [10]. For geometry when light is focused through a transparent substrate onto an interface with liquid, light-induced backside wet etching (LIBWE) is realized as originally developed for optical projection processing [21,22]. We adopted LIBWE for the direct laser writing. LIBWE delivers a better surface morphology of fs-laser processed areas in liquid. However, the surface quality (smoothness) after laser-assisted etching would suffer from loss in resolution and departure from the initial design due to chemical enhancement of material removal. Although debris can be removed in the liquid-assisted fs-laser machining, surface roughness is affected by the periodic structures [23].

In this paper, we report a systematical study of liquid assisted fs-laser machining of quartz plates in water, KOH, and HF solutions as explored in LIBWE geometry. The obtained micro-holes in liquid have a better morphology and a smaller diameter (1/3~1/2) compared with those fabricated in air. It has been found that the formation of laser induced periodic structures was decreased in solutions. Due to the redistribution of oxygen at the laser ablation site, chemical reactions affected formation of the induced periodic structures and provided better control over surface modification.

2. Materials and Methods

The schematics of the employed experimental laser microfabrication system are shown in Figure 1. A regeneratively amplified Ti:sapphire laser was used, which delivered pulses with a duration of 100 fs (FWHM), center wavelength of 800 nm, and repetition rate of 500 Hz. The diameter of the beam was approximately 6 mm, and then expanded to 18 mm with a beam expander, which was composed of two lenses with focal lengths of −50 mm and 150 mm. The laser beam was focused onto the upper surface (in contact with solution) through the sample using an objective lens (50× magnification, numerical aperture NA of 0.7 and aperture of 9 mm). The pulse energy was measured at the exit pupil of the objective lens. A quartz plate, 1-mm-thick, was mounted on a 3D translation stage with a positioning resolution of 0.1 μm (Figure 1). The upper surface was in contact with various solutions: deionized water, KOH (5% and 40% by mass ratio), and HF solution (2% or 5 mol/L). In this LIBWE case, the focused spot would not be influenced by bubble formation during laser fabrication. The laser fluence of the incident pulses was continuously tuned using a variable attenuator, while the laser pulse number was controlled by a shutter with a temporal resolution of 1 ms, which made it possible to obtain a single pulse.

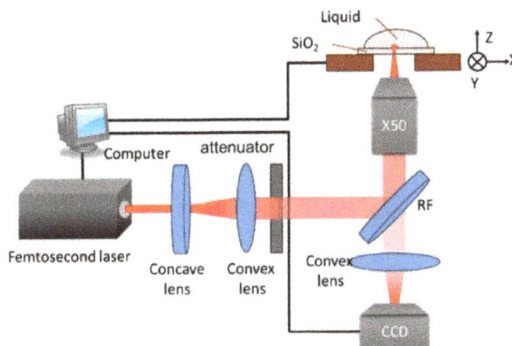

Figure 1. Schematics of the liquid-assisted femtosecond laser precision-machining system. The 50× magnification objective lens was uses, RF is the reflector. A CCD (charge-coupled device) was used for real-time observation of laser processing.

After laser fabrication, the sample was cleaned in an ultrasonic bath with KOH solution (40%, mass concentration) at room temperature, which removed all the debris from the surface thoroughly. After cleaning, the morphology of the ablated region was obtained with a scanning electron microscope (SEM, JSM-7500F, JEOL Ltd., Akishima-shi, Japan). At each machining parameter, a 5 × 5 hole-array was fabricated and all the results were the average of these 25 dots.

3. Results and Discussion

At first, the morphology of the holes fabricated with different laser pulse energies and pulse numbers were investigated after fabrication in air and deionized water. As shown in Figure 2, the laser pulse energy was set from 68 nJ to 107 nJ and the pulse number was changing from $N = 1$ to 100. Apparently, it indicates that the size of the holes ablated in deionized water was much smaller than those produced in air ambient (in LIBWE geometry). The interior of holes made in deionized water were relatively cleaner. In air, an unstructured ablation hole was observed at one pulse regardless of the pulse energy. Grating-like ripples with an orientation perpendicular to the laser polarization (the normal ripples) [24] were found around the irradiated regions at exposures above five pulses and 68 nJ/pulse energy. It is worth noting that a smaller ablation hole with diameter of about 120 nm was observed in the center of the irradiation region at 5–10 pulses of 68 nJ. There was still a pronounced ablation hole at the center for exposure by five pulses of 94 nJ and 107 nJ, which may be due to the competition between laser ablation and ripple formation as discussed later in more detail. When the pulse number was more than 10, laser induced ripples dominated the morphology regardless of the laser pulse energy.

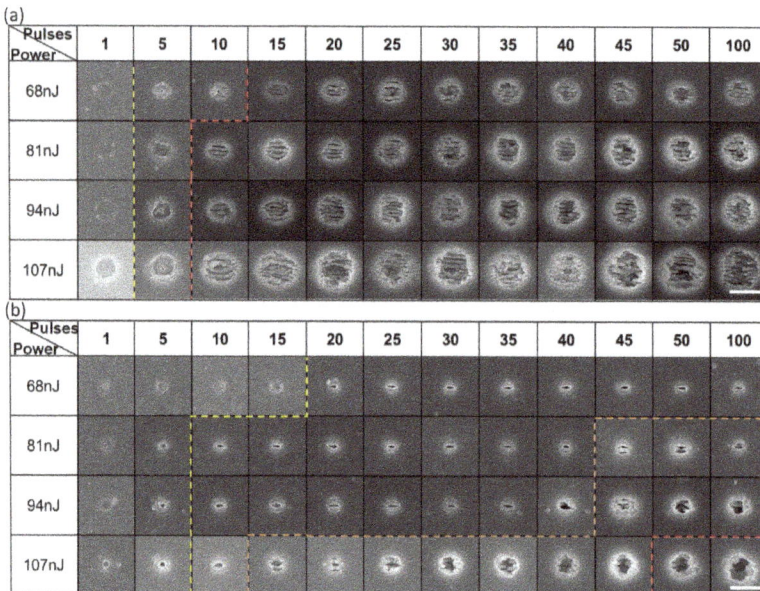

Figure 2. Scanning electron microscope (SEM) images of holes induced in air (**a**) and water (**b**) by different number of pulses varied from 1 to 100 and pulse energy ranging from 68 nJ to 107 nJ. Scale bars are 2 μm.

In water, the ablated products may interact and undergo chemical modifications in a short time after the laser pulse, especially inside the cavitation bubble [25]. Compared with air, the atom density of water is three orders of magnitude higher and confines the laser shock energy and plasma plume to a smaller region. As aforementioned (Figure 2a), ablation holes were observed with a smaller diameter after the first pulse regardless of the pulse energy. However, the hole formation by the ablation continued till 100 pulses at 68 nJ, 40 pulses at 81 nJ, 40 pulses at 94 nJ, and 10 pulses at 107 nJ and dominated laser structuring and material removal. Competition between direct hole formation by laser ablation and ripple formation was refined on larger areas and larger numbers of laser pulses. Ripples were printed directly on the surface as the ablation hole changed from circular, to elliptical, to linear and, finally, along the linear central feature (Figure 2b). This revealed a systematic change of surface modifications when all focusing conditions are same.

The diameter of the ablated area with the pulse number in air (a) and in water (b) is shown in Figure 3. In air, the diameter increased quickly with the pulse number below 30 pulses and then maintained saturation until 100 pulses. In water, the diameter changed much slower as the pulse increased at energies of 68 nJ, 81 nJ, and 94 nJ. However, it went up linearly with pulse accumulation at 107 nJ, which could also be observed in Figure 2b. The ablation threshold fluences of the quartz plate in air and water were evaluated by investigating the dependency of the ablated size (squared diameter, D^2) on the irradiation pulse energy. By fitting the data according to the equation (assuming Gaussian intensity profile) [26]

$$D^2 = 2\omega_0^2 \ln(E_p/E_{th}) \tag{1}$$

where ω_0 is the beam waist (radius) at the focal plane and E_{th} is the threshold energy. The $\ln(E_{th})$ could be obtained as asymptotic value when $D^2 \rightarrow 0$. The threshold fluences, ϕ_{th}, could be calculated for the peak amplitude

$$\phi_{th} = 2E_{th}/\pi\omega_0^2 \tag{2}$$

For a single shot, the calculated threshold pulse energies were about 46.4 nJ and 35.7 nJ in air and in water, respectively, which corresponded to the fluences of 2.3 J/cm² and 5.3 J/cm², respectively.

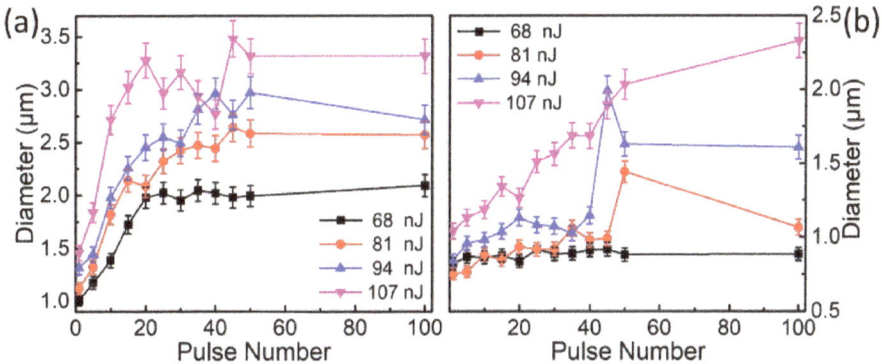

Figure 3. Diameter of the ablated area with pulse number in air (**a**) and water (**b**). Diameter is defined as a recognizable surface damage cross section in SEM images.

Laser induced periodic structures were found depressed in water as shown in Figure 2 with more details in Figure 4. Taking the period at 68 nJ, for example, the period observed at the ablation center at 5 or 10 pulses is around 150 nm. It quickly increased to around 260 nm at 30 pulses. Then it kept the saturation value up to 100 pulses. If the pulse energy increased, as shown in the Figure 4b, the center period at 50 pulses was around 289 nm at 81 nJ. While as a case in water, the period at 45 pulses and 68 nJ was around a slightly larger period of 300 nm. This is attributed to the stronger ablation which induced structures deep into the sub-surface volume. The induced nanostructures were formed by the accepted model of a standing wave at the interface of the active plasma (excited) layer and silica host (the refractive index around 1.48) rather than at the silica–air interface. The ambient refractive index has been demonstrated to be an important factor to the period of the induced structures; the smaller the refractive index, the larger the period [27,28].

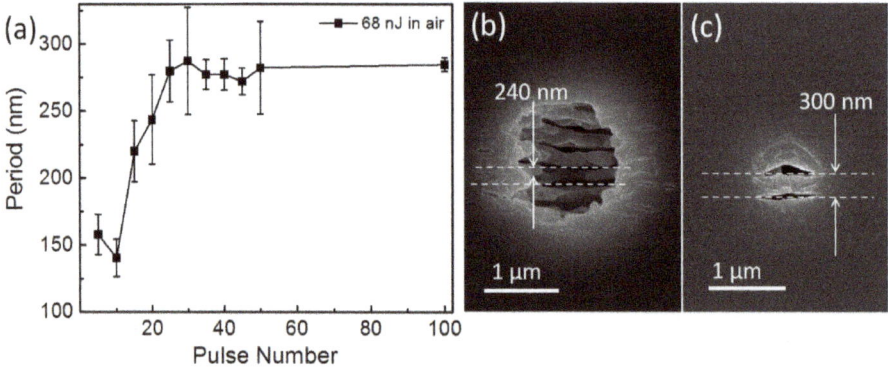

Figure 4. Laser induced periodic structures in air. (**a**) The period evolution at 68 nJ; (**b**) period at 50 pulses and 81 nJ in air; (**c**) period at 50 pulses and 81 nJ in water.

From the cross-section, it can be seen that the depth of periodic structures extended a micrometer into silica. Interestingly, the structures become squeezed with smaller periods at larger depths, expandable by the effective medium theory, i.e., a larger effective refractive index more far from the ablated interface.

The redistribution of the oxygen revealed lower concentration in the center but higher at the ablation border by energy disperse spectrum (EDS) measurement, as shown in Figure 5a. In the water, the morphology in cross section was different. The ablation hole without nanostructures extended deep into the bulk. However, the oxygen presented a similar distribution as that in the air ablated samples (Figure 5b). Ablation of silica evolves via building up of the electrostatic field between the fast removed electrons from the surface and lagging ions. High temperature and pressure conditions are created [29], which would cause phase changes and ion separation during the stage of hydrodynamic movement after an ultra-short laser pulse.

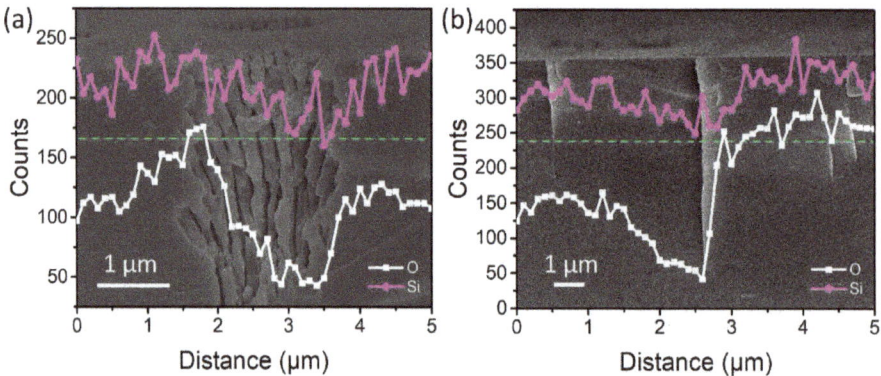

Figure 5. O and Si content along the depth of ablated surface (**a**) in air and (**b**) in water. The background SEM image has the same axial and lateral scales as the compositional plot.

The laser ablation process caused elemental redistribution, which is the most probable reason for selective etching of the ablated area and non-ablated area [30]. For example, the formed nanogratings in air would become different with larger nanovoids but could be made smooth after HF etching for 10 min, as shown in the second line in Figure 6. Slow etching at room temperature becomes accelerated by the ablation which is known also in LIBWE to cause cavitation (hence surface disintegration under negative tensile pressure). The ablation holes become rounder and deeper at the same pulse energies as compared with those in water (lines one and two in Figure 6).

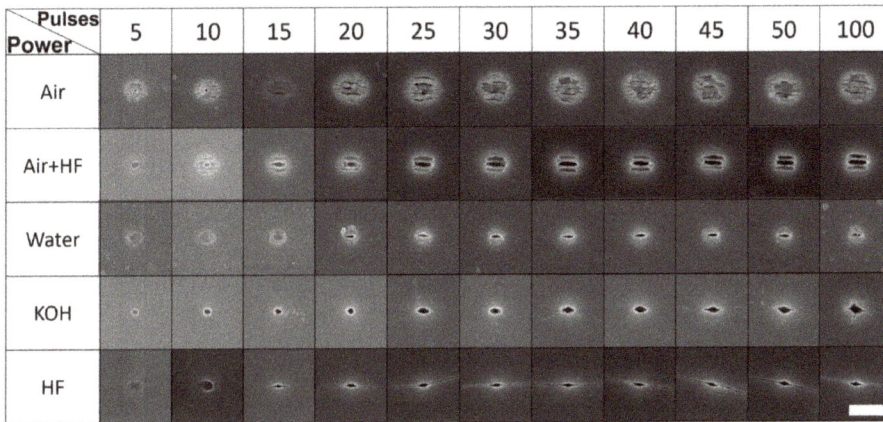

Pulses Power	5	10	15	20	25	30	35	40	45	50	100
Air											
Air+HF											
Water											
KOH											
HF											

Figure 6. Comparison between ablations in different solutions. The pulse energy for each is 68 nJ and the scale bar denotes 2 μm.

The morphologies of laser ablation in the etching solutions were very different for the HF and KOH: the surface becomes cleaner without laser induced ripples. Nanovoids of elliptical shape similar with those water but with a nanogaps extending perpendicular with the laser polarization were observed for KOH and HF. This can be attributed to the electronic heat conductivity enhancement along the electrical component of the oscillating light E-field which facilitates chemical reaction and thermal diffusion [31]. E-field control of electronic transport during laser irradiation (rather diffusional all directional spreading) is expected to manifest itself as enhanced chemical etching of material deposition and can be linked to the convectional mechanisms in liquid environment [32].

Femtosecond laser machining is discussed as 'cold processing' [33] due to having a smaller heat affected zone. However, this is achieved due to well-controlled energy deposition and high intensity with temperatures reaching more than $500\,°C$ [29] at the irradiated area, especially in the cavitation and bubble formation mode. High temperature enhances chemical reactions such as etching in KOH and HF solutions via the standard Arrhenius activation mechanism. The reaction in HF acid is more effective in material removal as compared with KOH since reaction product K_2SiO_3 is less dissolvable compared to SiF_4.

A unique feature of the laser ablated patterns is the well-centered hole formation especially recognizable with smaller numbers ($N = 5$) of pulses (see Figures 2 and 6). The pronounced central deeper ablation holes can be explained by the energy deposition. The strongest light absorption takes place at the regions where permittivity (epsilon) is near zero (ENZ) [34]. This occurs at the very central part and can explain the formation of stronger ablation which is localized into the sub-diffraction area. Since ENZ regions depend on the permittivities of the host as well as the free electron plasma, this feature of strong localization of ablation can be engineered and will be the focus of future studies.

4. Conclusions

In summary, we have systematically investigated the interaction of fs-laser pulses with fused quartz (silica) in air, water, KOH, and HF solutions in LIBWE geometry. Smaller holes and smooth surface were obtained in water and water solutions. The threshold in water had a similar decreasing trend with pulse accumulation. In addition, laser induced periodic structures were less pronounced in water. Etching enhancement along the E-field of a linearly polarized laser beam was observed in the laser ablation in solution, which is attributed to the thermal effect enhanced chemical reaction and E-field enhanced electronic conductivity. Our systematical investigation opens up prospects for a better controlled high precision nano-/micro-scale fabrication of hard and chemically inert materials. The mechanism of energy deposition into the ENZ regions is discussed.

Author Contributions: X.-W.C. performed the experiments; Q.-D.C. and H.-B.S. conceived and designed the experiments; S.J. analyzed the data; H.F. and L.Z. contributed materials/analysis tools; all the authors wrote this paper together.

Acknowledgments: This work was supported by the National Key R&D Program of China (2017YFB1104600), National Natural Science Foundation of China (NSFC) (61590930, 51335008, 61435005, 61605055), and the Program for JLU Science and Technology Innovative Research Team (JLUSTIRT) (2017TD-21).

Conflicts of Interest: The authors declare no conflict of interest.

References

1. Malinauskas, M.; Žukauskas, A.; Hasegawa, S.; Hayasaki, Y.; Mizeikis, V.; Buividas, R.; Juodkazis, S. Ultrafast Laser Processing of Materials: From Science to Industry. *Light Sci. Appl.* **2016**, *5*, e16133. [CrossRef]
2. Ni, J.; Wang, C.; Zhang, C.; Hu, Y.; Yang, L.; Lao, Z.; Xu, B.; Li, J.; Wu, D.; Chu, J. Three-Dimensional Chiral Microstructures Fabricated by Structured Optical Vortices in Isotropic Material. *Light Sci. Appl.* **2017**, *6*, e17011. [CrossRef]
3. Zhang, Y.L.; Chen, Q.D.; Xia, H.; Sun, H.B. Designable 3D Nanofabrication by Femtosecond Laser Direct Writing. *Nano Today* **2010**, *5*, 435–448. [CrossRef]
4. Juodkazis, S.; Mizeikis, V.; Matsuo, S.; Ueno, K.; Misawa, H. Three-Dimensional Micro- and Nano-Structuring of Materials by Tightly Focused Laser Radiation. *Bull. Chem. Soc. Jpn.* **2008**, *81*, 411–448. [CrossRef]
5. Guo, F.; Karl, A.; Xue, Q.-F.; Tam, K.C.; Forberich, K.; Brabec, C.J. The Fabrication of Color-Tunable Organic Light-Emitting Diode Displays via Solution Processing. *Light Sci. Appl.* **2017**, *6*, e17094. [CrossRef]
6. Guo, L.; Xia, H.; Fan, H.T.; Zhang, Y.L.; Chen, Q.D.; Zhang, T.; Sun, H.B. Femtosecond Laser Direct Patterning of Sensing Materials toward Flexible Integration of Micronanosensors. *Opt. Lett.* **2010**, *35*, 1695–1697. [CrossRef] [PubMed]
7. Tanaka, T.; Sun, H.B.; Kawata, S. Rapid Sub-Diffraction-Limit Laser Micro/Nanoprocessing in a Threshold Material System. *Appl. Phys. Lett.* **2002**, *80*, 312–314. [CrossRef]
8. Sugioka, K.; Cheng, Y. Ultrafast Lasers—Reliable Tools for Advanced Materials Processing. *Light Sci. Appl.* **2014**, *3*, e149. [CrossRef]
9. Shaheen, M.E.; Gagnon, J.E.; Fryer, B.J. Femtosecond Laser Ablation Behavior of Gold, Crystalline Silicon, and Fused Silica: A Comparative Study. *Laser Phys.* **2014**, *24*, 106102. [CrossRef]
10. Juodkazis, S.; Nishi, Y.; Misawa, H. Femtosecond Laser-Assisted Formation of Channels in Sapphire Using Koh Solution. *Phys. Status Solidi RRL* **2008**, *2*, 275–277. [CrossRef]
11. Ahsan, M.S.; Lee, M.S.; Hasan, M.K.; Noh, Y.-C.; Sohn, I.-B.; Ahmed, F.; Jun, M.B.G. Formation Mechanism of Self-Organized Nano-Ripples on Quartz Surface Using Femtosecond Laser Pulses. *Optik* **2015**, *126*, 5979–5983. [CrossRef]
12. Huang, M.; Zhao, F.; Cheng, Y.; Xu, N.; Xu, Z. Origin of Laser-Induced near-Subwavelength Ripples: Interference between Surface Plasmons and Incident Laser. *Acs Nano* **2009**, *3*, 4062–4070. [CrossRef] [PubMed]
13. Wang, L.; Chen, Q.-D.; Yang, R.; Xu, B.-B.; Wang, H.-Y.; Yang, H.; Huo, C.-S.; Sun, H.-B.; Tu, H.-L. Rapid Production of Large-Area Deep Sub-Wavelength Hybrid Structures by Femtosecond Laser Light-Field Tailoring. *Appl. Phys. Lett.* **2014**, *104*, 031904. [CrossRef]
14. Li, Q.K.; Yu, Y.H.; Wang, L.; Cao, X.W.; Liu, X.Q.; Sun, Y.L.; Chen, Q.D.; Duan, J.A.; Sun, H.B. Sapphire-Based Fresnel Zone Plate Fabricated by Femtosecond Laser Direct Writing and Wet Etching. *IEEE Photonics Technol. Lett.* **2016**, *28*, 1290–1293. [CrossRef]

15. Richter, S.; Heinrich, M.; Döring, S.; Tünnermann, A.; Nolte, S. Formation of Femtosecond Laser-Induced Nanogratings at High Repetition Rates. *Appl. Phys. A* **2011**, *104*, 503–507. [CrossRef]

16. Varkentina, N.; Dussauze, M.; Royon, A.; Ramme, M.; Petit, Y.; Canioni, L. High Repetition Rate Femtosecond Laser Irradiation of Fused Silica Studied by Raman Spectroscopy. *Opt. Mater. Express* **2015**, *6*, 79–90. [CrossRef]

17. Li, Q.K.; Cao, J.J.; Yu, Y.H.; Wang, L.; Sun, Y.L.; Chen, Q.D.; Sun, H.B. Fabrication of an Anti-Reflective Microstructure on Sapphire by Femtosecond Laser Direct Writing. *Opt. Lett.* **2017**, *42*, 543–546. [CrossRef] [PubMed]

18. Liu, X.Q.; Chen, Q.D.; Guan, K.M.; Ma, Z.C.; Yu, Y.H.; Li, Q.K.; Tian, Z.N.; Sun, H.B. Dry-Etching-Assisted Femtosecond Laser Machining. *Laser Photonics Rev.* **2017**, *11*, 1600115. [CrossRef]

19. Kaakkunen, J.J.J.; Silvennoinen, M.; Paivasaari, K.; Vahimaa, P. Water-Assisted Femtosecond Laser Pulse Ablation of High Aspect Ratio Holes. *Phys. Procedia* **2011**, *12*, 89–93. [CrossRef]

20. Li, Y.; Qu, S.L. Water-Assisted Femtosecond Laser Ablation for Fabricating Three-Dimensional Microfluidic Chips. *Curr. Appl. Phys.* **2013**, *13*, 1292–1295. [CrossRef]

21. Wang, J.; Niino, H.; Yabe, A. One-Step Microfabrication of Fused Silica by Laser Ablation of an Organic Solution. *Appl. Phys. A Mater.* **1999**, *68*, 111–113. [CrossRef]

22. Böhme, R.; Braun, A.; Zimmer, K. Backside Etching of UV-Transparent Materials at the Interface to Liquids. *Appl. Surf. Sci.* **2002**, *186*, 276–281. [CrossRef]

23. Ameer-Beg, S.; Perrie, W.; Rathbone, S.; Wright, J.; Weaver, W.; Champoux, H. Femtosecond Laser Microstructuring of Materials. *Appl. Surf. Sci.* **1998**, *127*, 875–880. [CrossRef]

24. Buividas, R.; Mikutis, M.; Juodkazis, S. Surface and Bulk Structuring of Materials by Ripples with Long and Short Laser Pulses: Recent Advances. *Prog. Quantum Electron.* **2014**, *38*, 119–156. [CrossRef]

25. Yan, Z.J.; Chrisey, D.B. Pulsed Laser Ablation in Liquid for Micro-/Nanostructure Generation. *J. Photochem. Photobiol. C* **2012**, *13*, 204–223. [CrossRef]

26. Baudach, S.; Bonse, J.; Krüger, J.; Kautek, W. Ultrashort Pulse Laser Ablation of Polycarbonate and Polymethylmethacrylate. *Appl. Surf. Sci.* **2000**, *154*, 555–560. [CrossRef]

27. Wang, L.; Xu, B.-B.; Cao, X.-W.; Li, Q.-K.; Tian, W.-J.; Chen, Q.-D.; Juodkazis, S.; Sun, H.-B. Competition between Subwavelength and Deep-Subwavelength Structures Ablated by Ultrashort Laser Pulses. *Optica* **2017**, *4*, 637–642. [CrossRef]

28. Wang, L.; Chen, Q.-D.; Cao, X.-W.; Buividas, R.; Wang, X.; Juodkazis, S.; Sun, H.-B. Plasmonic Nano-Printing: Large-Area Nanoscale Energy Deposition for Efficient Surface Texturing. *Light Sci. Appl.* **2017**, *6*, e17112. [CrossRef]

29. Chen, C.; Sun, H.B.; Guo, J.C.; Wang, L.; Chen, Q.D.; Yang, R.; Yu, Y.S. Monitoring Thermal Effect in Femtosecond Laser Interaction with Glass by Fiber Bragg Grating. *J. Lightw. Technol.* **2011**, *29*, 2126–2130. [CrossRef]

30. Juodkazis, S.; Yamasaki, K.; Mizeikis, V.; Matsuo, S.; Misawa, H. Formation of Embedded Patterns in Glasses Using Femtosecond Irradiation. *Appl. Phys. A Mater.* **2004**, *79*, 1549–1553. [CrossRef]

31. Rekštytė, S.; Jonavičius, T.; Gailevičius, D.; Malinauskas, M.; Mizeikis, V.; Gamaly, E.G.; Juodkazis, S. Nanoscale Precision of 3D Polymerization via Polarization Control. *Adv. Opt. Mater.* **2016**, *4*, 1209–1214. [CrossRef]

32. Louchev, O.A.; Juodkazis, S.; Murazawa, N.; Wada, S.; Misawa, H. Coupled Laser Molecular Trapping, Cluster Assembly, and Deposition Fed by Laser-Induced Marangoni Convection. *Opt. Express* **2008**, *16*, 5673–5680. [CrossRef] [PubMed]

33. Joglekar, A.P.; Liu, H.-H.; Meyhöfer, E.; Mourou, G.; Hunt, A.J. Optics at Critical Intensity: Applications to Nanomorphing. *Proc. Natl. Acad. Sci. USA* **2004**, *101*, 5856–5861. [CrossRef] [PubMed]

34. Gamaly, E.G.; Rode, A.V. Ultrafast Re-Structuring of the Electronic Landscape of Transparent Dielectrics: New Material States (Die-Met). *Appl. Phys. A* **2018**, *124*, 278. [CrossRef]

nanomaterials

MDPI

Article

Spontaneous Shape Alteration and Size Separation of Surfactant-Free Silver Particles Synthesized by Laser Ablation in Acetone during Long-Period Storage

Dongshi Zhang [1,†], Wonsuk Choi [1,2,3,†], Jurij Jakobi [4], Mark-Robert Kalus [4], Stephan Barcikowski [4], Sung-Hak Cho [2,5] and Koji Sugioka [1,*]

[1] RIKEN Center for Advanced Photonics, 2-1 Hirosawa, Wako, Saitama 351-0198, Japan; dongshi17@126.com (D.Z.); cws@kimm.re.kr (W.C.)
[2] Department of Nano-Mechatronics, Korea University of Science and Technology (UST), 217 Gajeong-Ro, Yuseong-Gu, Daejeon 34113, Korea; shcho@kimm.re.kr
[3] Department of Nano-Manufacturing Technology, Korea Institute of Machinery and Material (KIMM), 156 Gajeongbuk-Ro, Yuseong-Gu, Daejeon 34103, Korea
[4] Technical Chemistry I and Center for Nanointegration Duisburg-Essen (CENIDE), University of Duisburg-Essen, Universitaetsstrasse 7, 45141 Essen, Germany; jurij.jakobi@uni-due.de (J.J.); mark-robert.kalus@uni-due.de (M.-R.K.); stephan.barcikowski@uni-due.de (S.B.)
[5] Department of Laser & Electron Beam Application, Korea Institute of Machinery and Material (KIMM), 156 Gajeongbuk-Ro, Yuseong-Gu, Daejeon 34103, Korea
*   Correspondence: ksugioka@riken.jp; Tel.: +81-(0)48-467-9495
†   These authors contributed equally to this work.

Received: 5 July 2018; Accepted: 11 July 2018; Published: 13 July 2018

Abstract: The technique of laser ablation in liquids (LAL) has already demonstrated its flexibility and capability for the synthesis of a large variety of surfactant-free nanomaterials with a high purity. However, high purity can cause trouble for nanomaterial synthesis, because active high-purity particles can spontaneously grow into different nanocrystals, which makes it difficult to accurately tailor the size and shape of the synthesized nanomaterials. Therefore, a series of questions arise with regards to whether particle growth occurs during colloid storage, how large the particle size increases to, and into which shape the particles evolve. To obtain answers to these questions, here, Ag particles that are synthesized by femtosecond (fs) laser ablation of Ag in acetone are used as precursors to witness the spontaneous growth behavior of the LAL-generated surfactant-free Ag dots (2–10 nm) into different polygonal particles (5–50 nm), and the spontaneous size separation phenomenon by the carbon-encapsulation induced precipitation of large particles, after six months of colloid storage. The colloids obtained by LAL at a higher power (600 mW) possess a greater ability and higher efficiency to yield colloids with sizes of <40 nm than the colloids obtained at lower power (300 mW), because of the generation of a larger amount of carbon 'captors' by the decomposition of acetone and the stronger particle fragmentation. Both the size increase and the shape alteration lead to a redshift of the surface plasmon resonance (SPR) band of the Ag colloid from 404 nm to 414 nm, after storage. The Fourier transform infrared spectroscopy (FTIR) analysis shows that the Ag particles are conjugated with COO– and OH– groups, both of which may lead to the growth of polygonal particles. The CO and CO_2 molecules are adsorbed on the particle surfaces to form $Ag(CO)_x$ and $Ag(CO_2)_x$ complexes. Complementary nanosecond LAL experiments confirmed that the particle growth was inherent to LAL in acetone, and independent of pulse duration, although some differences in the final particle sizes were observed. The nanosecond-LAL yields monomodal colloids, whereas the size-separated, initially bimodal colloids from the fs-LAL provide a higher fraction of very small particles that are <5 nm. The spontaneous growth of the LAL-generated metallic particles presented in this work should arouse the special attention of academia, especially regarding the detailed discussion on how long the colloids can be preserved for particle characterization and applications, without causing a mismatch between the colloid properties and their performance.

The spontaneous size separation phenomenon may help researchers to realize a more reproducible synthesis for small metallic colloids, without concern for the generation of large particles.

Keywords: laser ablation in liquids; particle growth; carbon encapsulation; polygonal particle; core-shell particle; surfactant-free

---

1. Introduction

Laser ablation in liquids (LAL) is increasingly gaining worldwide attention as a newly emerging bottom-up and top-down combined technique for novel nanomaterial synthesis, offering diverse applications including optics and biology [1–4]. In particular, the LAL-synthesized nanomaterials possess a higher purity than the counterparts synthesized by wet-chemistry techniques [1], because surfactants or stabilizers are not required for colloid synthesis, which makes them very attractive and competitive for catalytic applications [5–8]. High-purity nanomaterial means that a large amount of active sites are exposed to the surrounding reactants or nanoparticles (NPs), which is actually a double-edged sword in nanoscience [7]. This is because spontaneous particle growth [9–12] may occur, which can increase the difficulty in accurately controlling the sizes [13–16] and the shapes [17–21] of the as-prepared nanomaterials.

Six mechanisms have been proposed for the growth of the LAL-generated nanomaterials, including LaMer-like growth [9], coalescence [22], Ostwald ripening [10], particle (oriented) attachment [18,23–25], adsorbate-induced growth [18,26], and reaction-induced growth [17]. Unlike the wet-chemistry synthesis method, where particle growth terminates when the seed concentration drops below the critical concentration [27], the particle growth can be sustained during the entire period of LAL, because every pulse ablation produces new seeds for particle growth. This is the reason novel football-like AgGe microspheres [17], with sizes up to 7 μm, can be obtained by LAL. Another difference from the conventional wet-chemistry techniques is the possibility of inducing multiple growth processes to generate different nanostructures. For example, large (>100 nm) hollow Mn_3O_4 spheres are formed from ca. 20 nm cubic particles that are assembled from smaller 5–8 nm smaller particles after a long-period LAL [18]. Previously, the investigation on particle growth has mainly been on the fragmentation of large metallic particles, such as Pt [10] and Au [11] into 2~3 nm particles, and then their evolution over time is observed. No attempt has been made to study the growth of the metallic particles just after laser irradiation, which occupies the dominant position in all three liquid-phase laser synthesis methodologies, LAL, laser fragmentation in liquids (LFL), and laser melting in liquids (LML) [1].

LAL is currently most frequently implemented in a batch chamber where the ablation is often accompanied by LFL when the particles enter the beam path in the liquids [1]. The fragmentation behavior of the particles may become dominant when implementing the LAL for a long period. Long-period LAL means that the LAL should last more than half an hour, in order to generate highly saturated colloids in a limited volume of liquids [9]. In this case, a large amount of active surfactant-free particles are generated by LFL, which increases the chance for the particles to encounter each other by Brownian motion, and then to grow overtime during storage, transportation, and when employed in applications. If growth does occur, then it should attract special attention from academia, because until now, the characterization and application of metallic particles such as Ag NPs [28–31] have been seldom described in the literature. Therefore, it is not clear whether the size information shown in the literature has been obtained from either freshly prepared colloids or after a certain period of growth. The particle growth of the metallic particles is also accompanied by the shape transformation into nonspherical nanostructures, such as nanowires [10,11]. Tsuji's group reported that the post laser irradiation of LAL-generated Ag colloids in either a citrate [32], polyvinylpyrrolidone (PVP) aqueous solution [33], water [34], or acetone-water mixed solution [35] may cause the formation of nanoplates and nanorods during colloidal aging [36]. However, it is still unknown whether the Ag particles synthesized by long-period LAL can still grow into polygonal products, and the maximal size at which the particle growth terminates is also unknown, especially in the case of LAL in organic solvents, where the solvent decomposition definitely occurs to generate a large amount of carbon or hydrocarbons, which may precipitate on the active particles to terminate the particle growth.

In this paper, one hour long-period LAL of Ag in acetone, using a femtosecond laser (fs-LAL), was implemented to investigate the particle growth behavior. Laser powers of 300 mW and 600 mW were used to change the productivity of the Ag colloids, as well as the amount of both ultra-small particles and large particles with sizes less than and greater than 10 nm, respectively. The variation in the surface plasmon resonance (SPR) band of the Ag colloids during storage was recorded using UV-VIS spectroscopy, while the particles' sizes before and after the six month storage were analyzed using transmission electron microscopy (TEM), both of which provide evidence for particle growth and newly discovered size separation phenomena. The surface chemistry of the Ag NPs was analyzed using Fourier transform infrared spectroscopy (FTIR), to identify the chemical groups that may be responsible for the particle growth. This study was complemented by nanosecond (ns) laser ablation to demonstrate that particle growth is not limited to fs-LAL. Finally, the scenarios for both the particle growth and the size separation phenomena are proposed.

2. Results and Discussion

2.1. Femtosecond Laser Ablation

It has been reported that the redshift of the SPR peaks for the Ag NPs is often indicative of an increase of the particle size [37] via either particle ripening or coalescence, thus giving indirect evidence for particle growth. In our experiments, the variations in the absorption spectra of the fresh Ag colloids synthesized by LAL at laser powers of 300 mW and 600 mW, were recorded within 32 h, with a time interval of 30 min, as shown in Figure 1a,b. As the aging time increases, in both cases, the colloidal SPR peaks redshift from 404 nm to 410 nm, which suggests the spontaneous particle growth during colloidal storage. The higher SPR peak intensity of Ag colloids synthesized at 600 mW compared with those synthesized at 300 mW, indicates that the LAL at a higher laser power gives rise to a slightly higher productivity of the colloids. A slight increase in the SPR peak intensity with aging is observed in both cases, which is attributed to the increased concentration of the Ag NPs, caused by the evaporation of acetone. After six months of storage, during which time the growth was already terminated, the absorption spectra of the colloids were characterized again, as shown in Figure 1c,d. It is clear that the evaporation of the acetone further increases the colloidal concentrations, even though the colloids were sealed with acetone in glass containers. To compensate for the evaporated acetone, the colloids were diluted with additional acetone, and characterized (pink and blue curves in Figure 1c). The magnified SPR peaks in the wavelength range of 360–500 nm show a further redshift to 414 nm in both cases, which indicates that particle growth should continue even after the 32 h of storage. The inset figures in Figure 1c show photographs of the colloids in acetone after six months of storage. Both of the Ag colloids are orange colored. The higher concentration of Ag colloids prepared at a higher power is evident from the darker orange color. These colors are different from the light yellow color of the Ag colloids reported previously [38], probably due to the much higher concentration of Ag colloids presented in this work. Particle precipitation was also observed at the bottom of the glass containers. Therefore, a subsequent analysis of both of the stable colloids in liquids, and the particles precipitated at the bottom of the glass containers, was performed.

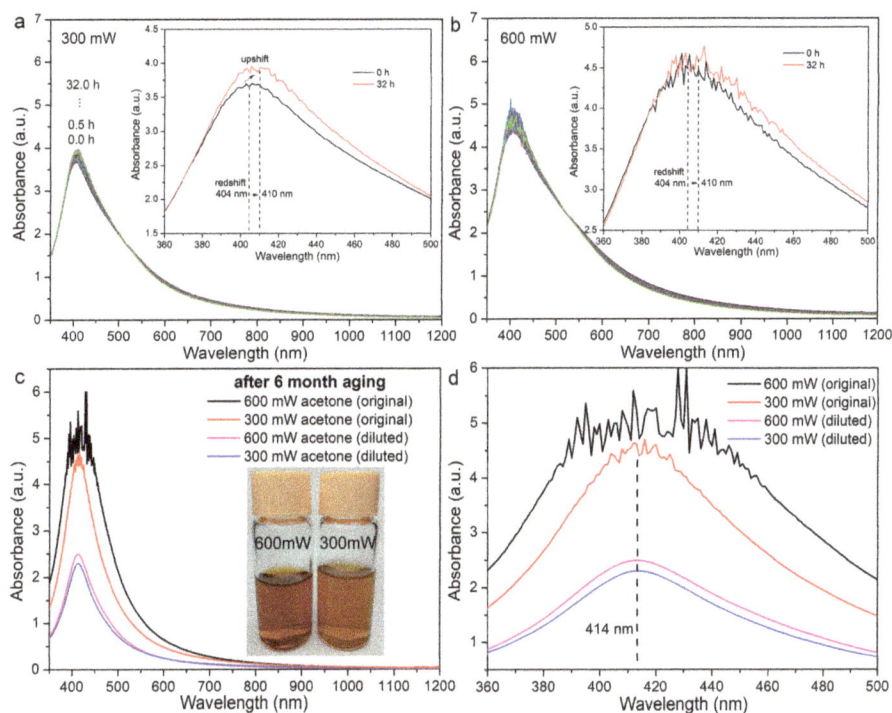

Figure 1. Absorption spectra for Ag nanoparticles (NPs) synthesized by laser ablation of Ag in acetone, and then stored for various periods. Upper: spectra for samples prepared at a laser power of (**a**) 300 mW and (**b**) 600 mW, and stored within 32 h (insets: enlarged spectra in the wavelength range of 360 nm to 500 nm for samples stored for 0 and 32 h). Lower: spectra for (**c**) samples prepared at laser powers of 300 mW and 600 mW, and stored for six months (inset: photos of aged Ag NPs in acetone), and (**d**) enlarged spectra in the wavelength range of 360 nm to 500 nm (black and red curves: as-aged colloidal samples; pink and blue curves: diluted as-aged samples prepared by the addition of pure acetone to the colloids).

Figures 2 and 3 show the TEM images and the calculated size distributions of the fresh and six month aged Ag particles generated by LAL, at the laser powers of 300 mW and 600 mW, respectively. In accordance with the redshift of the SPR peaks, shown in Figure 1a,b, particle growth indeed occurs (Figures 2a–f and 3a–f), especially for ultrasmall particles with sizes less than 10 nm. The average sizes of the fresh Ag particles synthesized at 300 mW and 600 mW were evaluated to be 5.9 ± 7.6 nm and 5.9 ± 12.2 nm (Figure 3g), respectively, which confirmed that the average particle size was almost independent of the laser power. However, a greater amount of particles larger than 10 nm were generated at a higher laser power of 600 mW (Figure 2a vs. Figure 3a), as indicated by the increased ratio between the big and small particles (Figure 2g vs. Figure 2h). The majority of the Ag NPs are in the range of 1–10 nm, occupying ca. 90% of the total amount for both of the cases. A further subdivision of the size distributions of the Ag NPs shows that most of the particles are ca. 2 nm (Figures 2c and 3c). After six months of storage, the average sizes of the Ag colloids increase to 7.4 ± 7.6 nm and 7.8 ± 8.2 nm, respectively. In the case of the colloids synthesized at 600 mW, a significant decrease in the large particles was observed (Figure 3a,d).

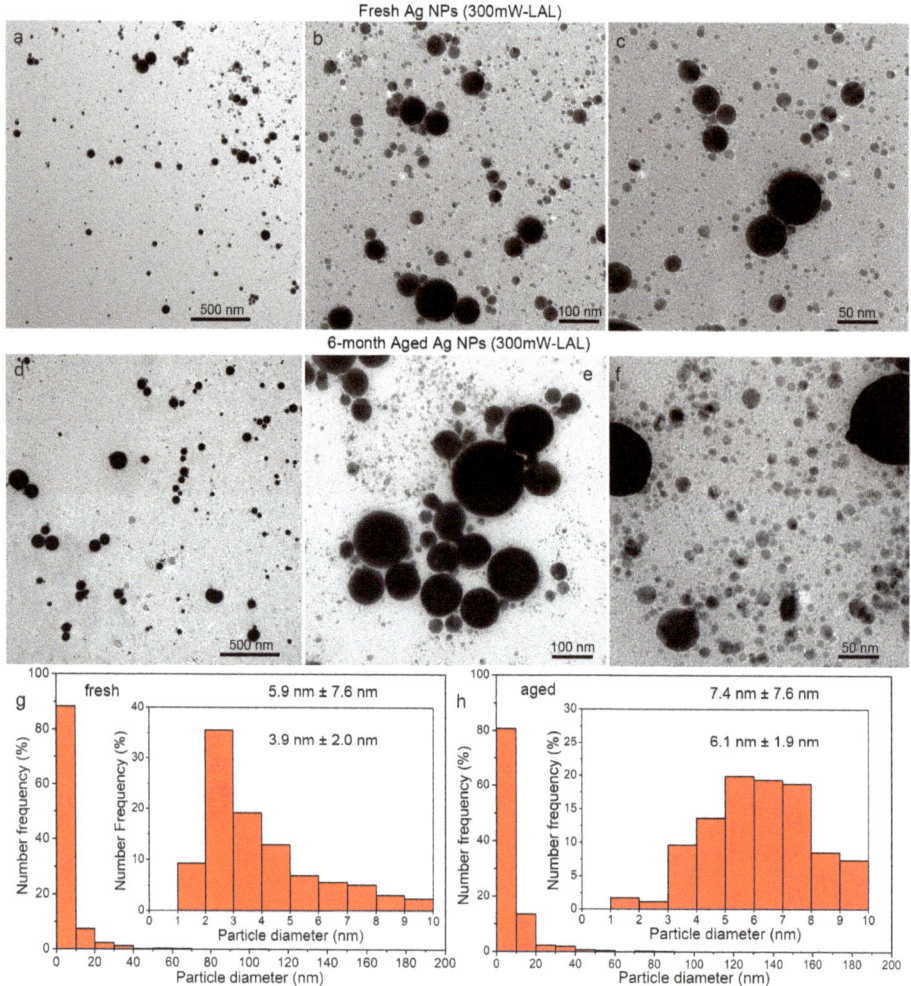

Figure 2. TEM images (**a–f**) and size distributions (**g,h**) of fresh (**a–c,g**) and six month (**d–f,h**) aged Ag particles, synthesized by laser ablation of Ag in acetone, at a laser power of 300 mW.

In both cases, the number ratios of the as-aged colloids with sizes in the range of 10–20 nm increased to 12–14% after 6 months of storage. The number ratios of the as-aged colloids with a diameter of 20~30 nm were almost the same as those of the fresh colloids (Figure 2g,h and Figure 3g,h). The number ratio of the as-aged colloid with a diameter of 30~40 nm increased slightly for the as-aged 300 mW LAL colloid, but decreased for the as-aged 600 mW LAL colloid. The number ratios of the Ag NPs with diameters larger than 40 nm were less than 1%, which is almost negligible for both cases, so it is difficult to quantify the variation in the number ratios of the colloids. However, a comparison of the particle morphologies of the fresh and as-aged colloids clearly shows that the number ratio of the particles larger than 40 nm did not change significantly (Figure 2a vs. Figure 2d) after six months of storage for the 300 mW LAL colloid, but did decrease significantly for the as-aged 600 mW LAL colloid (Figure 3a vs. Figure 3d).

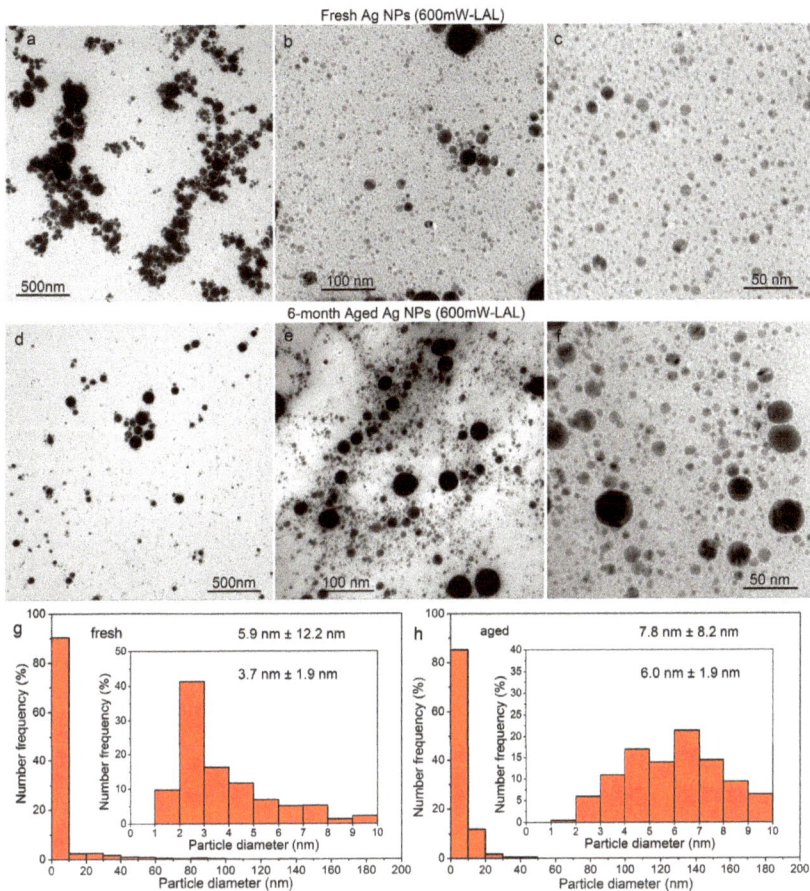

Figure 3. TEM images (**a–f**) and sizes distributions (**g,h**) of fresh (**a–c,g**) and six month (**d–f,h**) aged Ag particles, synthesized by laser ablation of Ag in acetone at laser power of 600 mW.

In more detailed observations of large particles, regardless of them being fresh (Figure 4a–c) or six months aged colloids (Figure 4d–f) prepared at laser powers of 300 mW and 600 mW, the carbon-encapsulated particles with core sizes ranging from 25 nm to ca. 200 nm, and a carbon shell thickness of 3~10 nm are all found, as shown in Figure 4. Some smaller particles were captured by the carbon shells, which led to the formation of Ag@C-Ag truffle-like aggregates. The crystalline structure of the Ag cores was confirmed by X-ray diffraction (XRD) characterization, as shown in Figure 5a. The featured peaks of the colloids fit well with the standard card for Ag (ICCD No. 03-065-2871). The presence of the carbon shells was verified by Raman spectroscopy observation of both the D-band (1360 cm^{-1}) and G-band (1582 cm^{-1}) peaks of carbon (Figure 5b). The ratio of the D-band to the G-band peak intensities was calculated to be 0.98. The G-band is associated with the ordered graphite (sp^2) structure, while the D-band is related to the disordered graphite layers, such as soot, chars, glassy carbon, and evaporated amorphous carbon [39]. Consequently, it can be concluded that the carbon shells possess high ratios of disorders. According to Robertson [39], the carbon shells that are composed of both crystalline graphite and amorphous carbon can be assigned to diamond-like carbon (DLC), a metastable form of carbon. Because of the presence of abundant hydrogen, which are generated from the LAL-induced acetone decomposition, it is highly possible that other products, such as hydrogenated amorphous carbon (a-C: H) and tetrahedral amorphous carbon (ta-C) [39] also form during LAL. Nevertheless, their existence cannot be confirmed by the Raman spectrum, shown in Figure 5b. The identification and quantification of different carbon disorders will be a focus of our future studies.

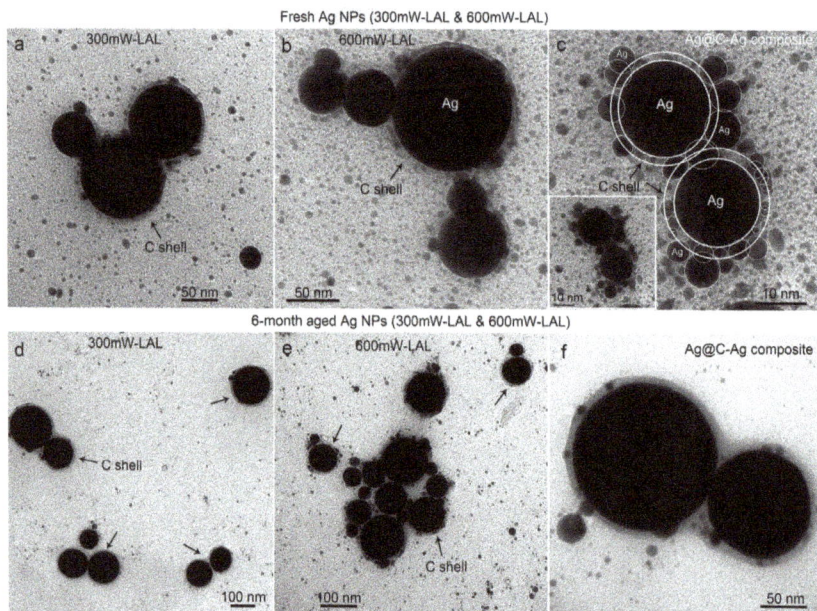

Figure 4. TEM images of the Ag@C core-shell particles observed in (**a–c**) fresh and (**d–f**) six month aged colloids synthesized by laser ablation in liquids (LAL), at laser powers of 300 mW (**a,d**) and 600 mW (**b,e**). The TEM images in (**c,f**) show the Ag@C-Ag composites from the fresh and the six month aged colloids.

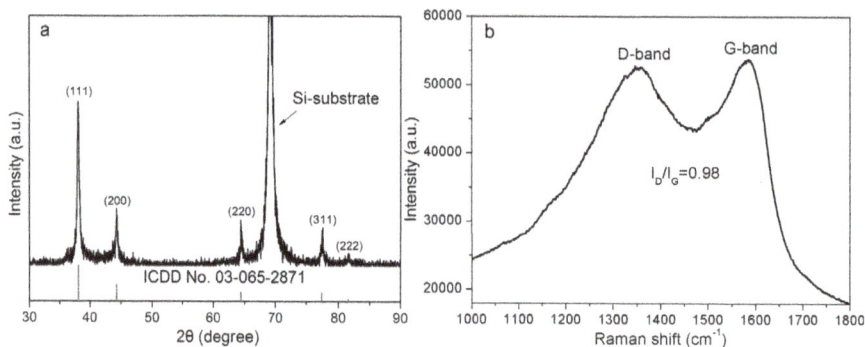

Figure 5. (**a**) XRD pattern and (**b**) Raman spectrum of the LAL-synthesized Ag NPs.

Some polygonal Ag nanocrystals, such as triangular (Figure 6d), pentagonal (Figure 6c,f), hexagonal (Figure 6a,d,f), octagonal (Figure 6b,e), and some spherical particles with sharp edges, (Figure 6f) were observed, of which the sizes were in the range of 5–50 nm, similar to the Ag nanocrystals prepared by the reduction of AgNO$_3$ [40]. The maximal sizes of the newly grown Ag nanocrystals are much smaller than the 100~500 nm crystals obtained by the post-irradiation of the

LAL-generated Ag spheres [33], which indicates that the surrounding carbon atoms may cover the newly formed nanocrystals and inhibit their further growth into larger particles. The following results provide evidence for the spontaneous growth of small particles into nonspherical Ag nanocrystals: (1) the amount of the ultrasmall particles, of 2–3 nm in diameter, is significantly decreased after storage (Figure 2g,h and Figure 3g,h); (2) polygonal particles are not surrounded by ultrasmall particles (Figure 6), unlike the spherical Ag NPs encapsulated by carbon shells (Figure 4); and (3) previous reports show that the size of the Ag NPs that correspond to an SPR band of 414 nm is around 30 nm [41], which is much larger than the ca. 8 nm (Figures 2h and 3h) of the six month aged Ag colloids with the same SPR position (Figure 1c,d) in this work. Considering that the polygonal Ag nanostructures, such as the hexagonal and triangular Ag nanoplates, often have a higher SPR band position than the Ag spheres with the same sizes [42], it is reasonable to deduce that the formation of the polygonal particles is the main contribution to the redshift of the SPR band during the storage of the colloids, rather than a change of the particle size. From the occurrence of particle growth, the building blocks (fresh ultrasmall Ag NPs) should be surfactant-free, which facilitates the attachment and ripening of ultrasmall particles into various polygonal nanoparticles, with the growth direction governed by the surface bindings [9,18], which will be discussed below.

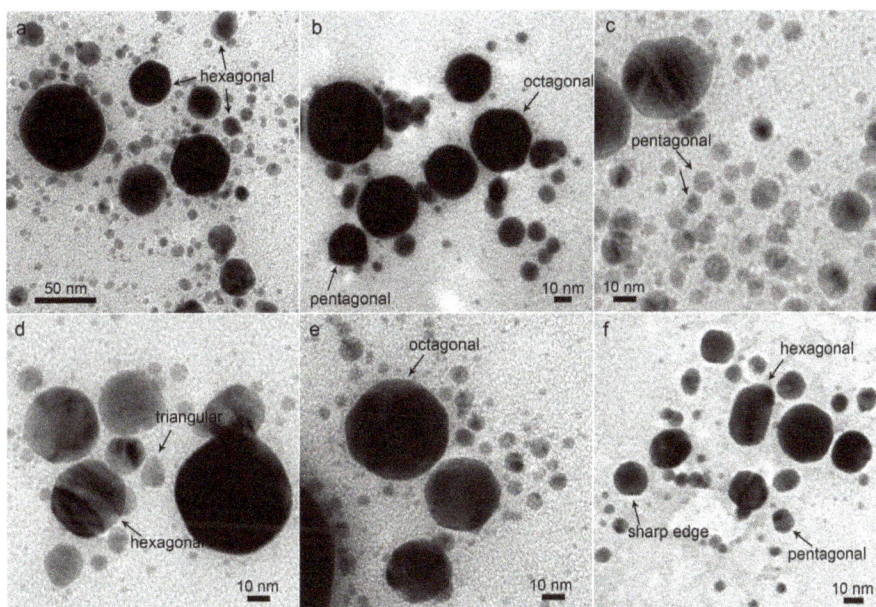

Figure 6. (a–f) Polygonal particles obtained after six months of storage of the colloids.

To determine the surface groups that may direct the particle growth into the polygonal particles, the six month aged Ag particles were analyzed using FTIR, and the results are shown in Figure 7. The vibration bands at 1165 cm^{-1} and 1250 cm^{-1} are assigned to the wagging and the rocking vibrations of CH_2, respectively, while the peaks at 1375 cm^{-1} and 1734 cm^{-1} are assigned to the stretching vibrations of C–H alkane [43] and C=O [44], respectively. The broad peak between 3000 cm^{-1} and 3500 cm^{-1} is because of the OH stretching [44]. The broad bands of 2061~2158 cm^{-1} and 2325~2343 cm^{-1} are related to the stretching vibrations of the adsorbed CO [45,46] and CO_2 [47] molecules, respectively. Therefore, the CO-metal [48] and CO_2-metal complexes [49], such as $Ag(CO)_n$ (n = 1~3) [50] and Ag–O–C–O, may be generated during LAL (Figure 7b). The three vibration peaks at 2902 cm^{-1}, 2932 cm^{-1}, and 2960 cm^{-1}, are due to the CH stretching from the alkyl groups [44].

The FTIR analysis indicates that the acetone molecules are decomposed into CO_2, CO, alkanes, and OH–groups during LAL, which may either strongly adsorb onto the Ag NPs or interact with the Ag NPs to form $Ag(CO)_n$ or $Ag(CO_2)_n$ complexes. As a result, the colloids are endowed with ultrahigh stability, which enables them to self-stabilize, even after six months of storage. Tsuji's group confirmed that shape transformation from the Ag spheres into prisms is not induced by acetone molecules, but should be caused by water molecules. In the present case, considering the absence of water molecules in the liquids, the OH– formed by the acetone decomposition and reconstruction is one possible candidate to cause the anisotropic growth of Ag particles [35] via the Ostwald ripening process [33]. Chemical adsorbed carboxylate (COO–) groups may be another candidate to trigger the selective Ag crystal facet growth during ripening [51].

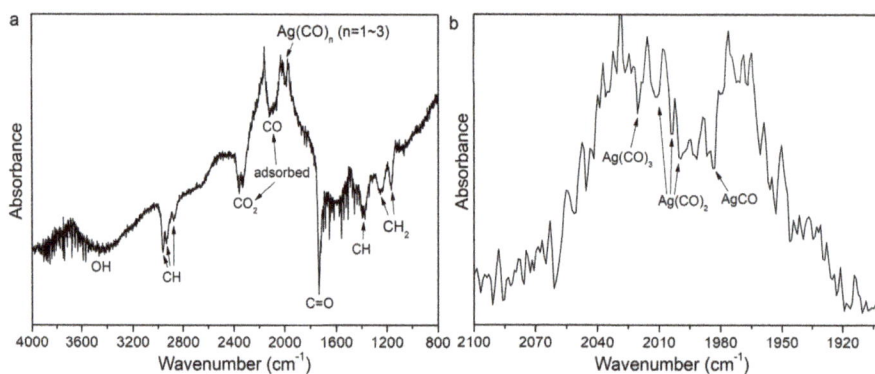

Figure 7. (**a**) FTIR spectrum of the six month aged Ag NPs and the (**b**) magnified spectrum where the peaks of $Ag(CO)_n$ (n = 1–3) complexes are located.

The optical images of the six month aged colloids show that some particles have already precipitated during storage (inset images in Figure 8c,f). In comparison with the particle morphologies of the fresh and six month aged colloids synthesized by LAL at a laser power of 600 mW (Figure 3a,d), it is also clear that the amount of particles with diameters larger than 40 nm decreased dramatically after a long-period storage (Figure 3g,h), which is due to the precipitation of larger particles. Figure 8 shows the TEM images of the precipitated particles prepared at laser powers of 300 mW and 600 mW after six months of storage. The precipitated particles are mainly particles larger than 40 nm, which are encapsulated by carbon to form particle networks, which also encapsulate a certain amount of particles with sizes less than 40 nm, so that the number ratios of the ultrasmall colloids (1–10 nm) decrease (Figures 2h and 3h). Despite the precipitation induced by the carbon capture, the number ratios of the colloids with sizes of 10–20 nm still increase, and the number ratios of the colloids with sizes of 20–30 nm remain almost unchanged. This means that the continuous supplement of the Ag NPs with sizes of 10–30 nm as a result of the particle growth of the ultrasmall surfactant-free Ag NPs occurs. An analysis of the change in the colloidal size distribution after six months of storage (Figure 2g,h and Figure 3g,h) led us to conclude that the colloids synthesized by LAL at higher powers possess a higher ability to separate the Ag colloids with sizes less than 40 nm. This is because more carbonaceous substances are generated at 600 mW, and these can then easily capture the large particles to form aggregate and precipitate over time, whereas the 300 mW LAL generates fewer carbon clusters, so that the size separation ability decreases, making it less efficient to separate the particles larger than 40 nm. The carbon 'captors' that originate from the LAL-induced decomposition of acetone have three forms, the carbon shells of Ag@C particles that capture small Ag particles to form truffle-like aggregates (Figures 4c and 8c), the active wandering carbon clusters that gradually precipitate on particles to induce the formation of particle networks (Figure 8b,e), and the carbon nanosheets (Figure 9) that mainly capture the ultrasmall Ag particles with sizes less than 10 nm (Figure 9d–f). Recently, Escobar-Alarcón showed that the LAL of graphite in water could produce carbon nanosheets [52], in which they proposed that graphene exfoliation was the formation mechanism for carbon nanosheets. However, in our case, carbon comes from the decomposition of acetone molecules during LAL. Hence, the carbon nanosheets should form by the self-assembly of carbon clusters inside liquids, which can also explain the irregular structure shapes of nanosheets shown in Figure 9 and in Escobar-Alarcón et al. [52].

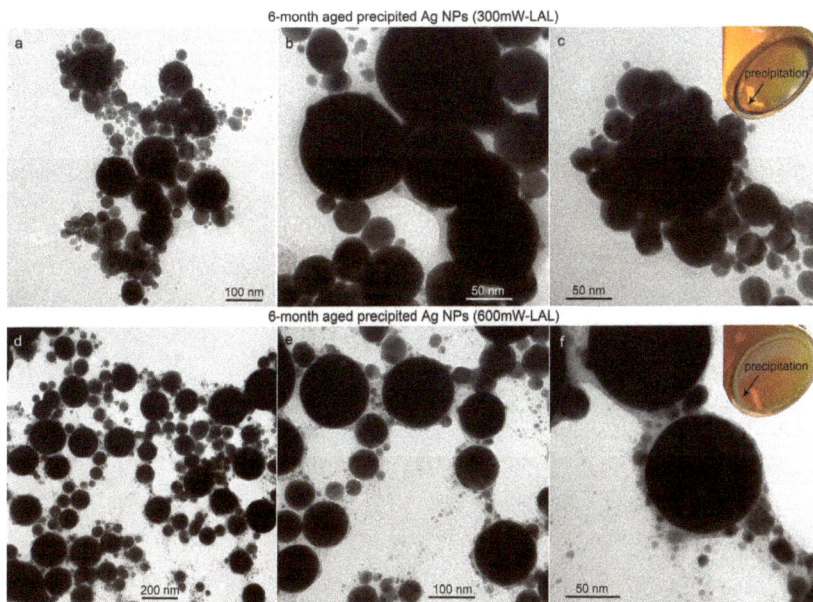

Figure 8. Precipitated particles after the six month storage of the Ag colloids, synthesized by femtosecond LAL at laser powers of (**a–c**) 300 mW and (**d–f**) 600 mW, respectively. The arrows shown in the optical images in the insets (**c,f**) indicate the particle precipitation after six months storage.

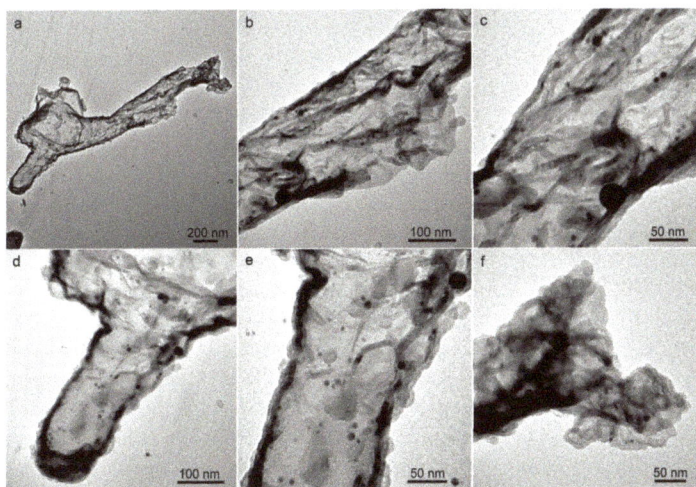

Figure 9. (**a–f**) TEM images of the carbon nanosheet formed during storage of the colloids.

Overall, the particle growth followed by the size separation phenomena was determined to occur during storage. The scenario is summarized in Figure 10. After long-period LAL, a large amount of Ag (orange colored spheres) and C (black colored spheres) clusters together to generate large Ag@C core-shell particles (Figure 10a). During the colloidal storage, the growth of ultrasmall surfactant-free Ag particles of 1–25 nm in diameter occurs through the Ostwald ripening mechanism (Figure 10b,c). Meanwhile, the soft carbon shells of large particles capture the surrounding small particles to form Ag@C-Ag aggregates. The aggregated particles precipitate at the bottom of the container (Figure 10b,c). The precipitation of the large particles cause the size distribution of the Ag particles in the supernatant to narrow by carbon encapsulation, which is termed size separation in this work. The self-size separation, with the aid of the large particle precipitation by carbon-encapsulation, can offset the disadvantage of the increased amount of large particles that are formed at a higher laser power, thus allowing a higher-productivity synthesis of metallic colloids with the uniform size distribution with that synthesized by the LAL at a lower laser power. The sediments may be technically separated by centrifugation or filtering to yield stable, monomodal Ag colloids in acetone.

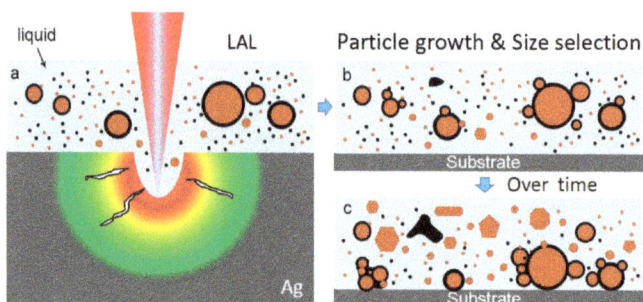

Figure 10. (**a–c**) Schematic of particle growth into polygonal nanocrystals, followed by the size separation of ultrasmall particles by large particle precipitation, due to carbon encapsulation. In the liquids, Ag and carbon are denoted by orange and black colored spheres, respectively.

2.2. Nanosecond Laser Ablation

Nanosecond laser ablation of Ag was performed in acetone for comparison, so as to investigate whether the spontaneous size separation and growth evolution into the polygonal crystals are specific for the fs-LAL, or whether they occur at a significantly longer pulse duration. Figure 11 shows the absorbance spectrum of the Ag NPs synthesized by the ns laser ablation of Ag in acetone, and that of the colloid after storage for five weeks. The SPR position of the Ag colloids was slightly redshifted from 400 nm to 407 nm after five weeks of storage, accompanied by the broadening of the SPR peak. Both indicate the change of the colloidal properties. Figure 12 shows the morphologies of the fresh and the aged Ag NPs, and their size distributions. Compared to the fresh Ag colloids with an average size of 8.93 ± 2.7 nm, the Ag NPs size was slightly increased to 11.1 ± 3.9 nm after storage for five weeks. Regarding the ultrasmall Ag NPs of 1–10 nm, the average size is increased from 7.5 ± 1.5 nm to 8.2 ± 1.3 nm after long-term storage. A significant decrease in the number ratio of the 1–10 nm particles from 68% to 47%, and the dramatic increase in the number ratio of the 10–20 nm particles from 22% to 50% (Figure 12g,h), provide evidence for the gradual growth of hte ns laser-synthesized Ag NPs in acetone during storage. The observation of more polygonal nanocrystals from the five week aged Ag colloid (Figure 12f) compared with the fresh colloid (Figure 12c) indicates that particle growth is always accompanied by the shape alteration of the metallic particles, regardless of the pulse duration used for colloid synthesis. Particle growth and simultaneous shape alternation are considered to constitute the main reason for the redshift and broadening of the SPR band (Figure 11).

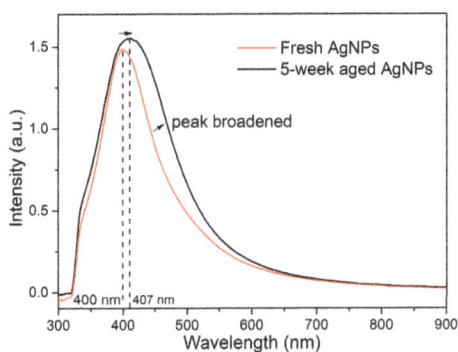

Figure 11. Absorption spectra for Ag NPs synthesized by ns laser ablation of Ag in acetone and then stored for five weeks.

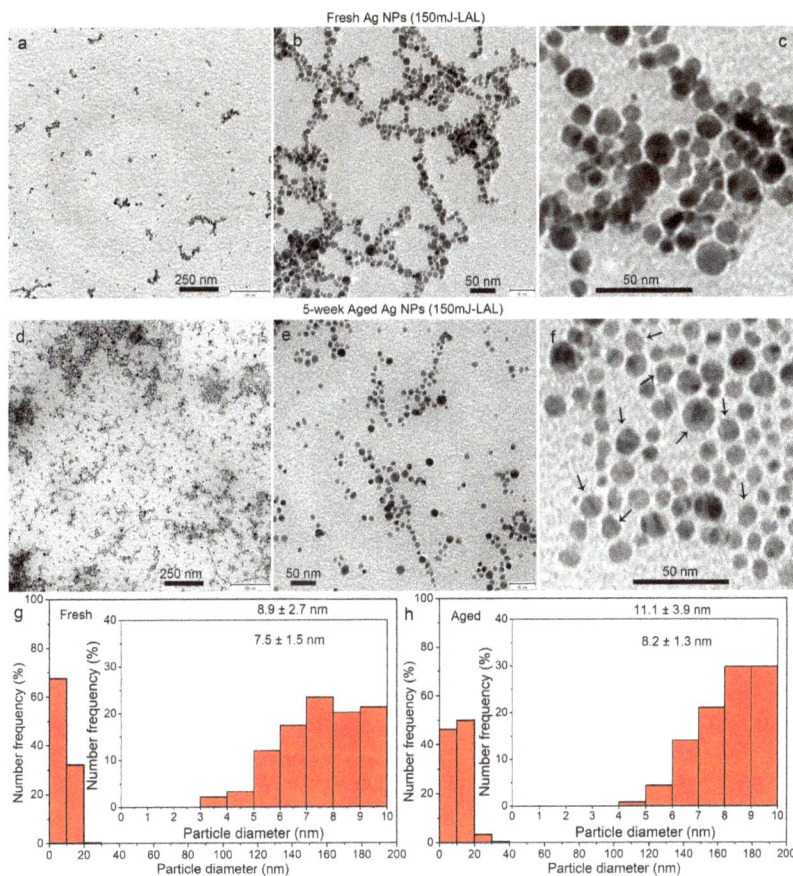

Figure 12. TEM images (**a–f**) and sizes distributions (**g,h**) of fresh (**a–c,g**) and five week (**d–f,h**) aged Ag particles synthesized by ns laser ablation of Ag in acetone at a pulse energy of 150 mJ.

However, compared to those particles synthesized by fs-LAL (Figures 2a–f and 3a–f), the carbon shells were almost negligible for the ns-LAL generated Ag NPs (Figure 12a–f). No big particles larger than 40 nm were observed, which indicates that the fs-LAL, at high fluences (170 and 191 J/cm^2 for 300 mW and 600 mW, respectively), causes more severe degradation of the acetone molecules than the ns-LAL at a low laser fluence (3 J/cm^2). An FTIR study on the LAL in tetrahydrofuran showed that the carbonylic compounds were dominantly generated during ns-LAL, whereas the more olefinic species were dominantly generated during fs and picosecond (ps) laser ablation; therefore, the fs-LAL created more hydrophobic species [53]. If such hydrophobic species are also preferably formed during fs-LAL in acetone, then this could explain why their lower solubility triggers phase separation (into soft carbon shells, adsorbed on hydrophobic metal Ag), which supports the aggregation and sedimentation processes.

Regarding the polydispersity of the colloid, fs-LAL at high intensities leads to the formation of large Ag droplets jetting off the molten metal layer. This jetting is caused by Rayleigh instability, with the droplets evident even in front of an expanding cavitation bubble boundary [54]. These droplets solidify as large particles, so that the size distributions from the fs-LAL show stronger bimodality (even large spheres) compared to the ns-LAL.

Therefore, the ns-LAL in acetone yields colloids that are far less polydisperse than the fs-LAL. However, after the sedimentation or size separation of the aged fraction, the final colloid in the supernatant of the fs-LAL colloid has significantly smaller primary particles. Comparing the number ratios of 1–10 nm ultrasmall particles (80–90% for fs-LAL, Figures 2g and 3g, vs. 68% for ns-LAL, Figure 12g), it can be easily deduced that fs-LAL is more efficient than ns-LAL for the generation of ultrasmall particles, at the expense of the mass yield lost during the size separation. In particular, the histograms from the TEM measurements show that the fs-LAL supernatant will contain a significantly larger portion of very small particles (<5 nm).

3. Materials and Methods

Silver colloids were synthesized by laser ablation of Ag in acetone by fs laser (FGPA μ Jewel D-1000-UG3, IMRA America Inc., Ann Arbor, MI, USA), with a pulse duration, wavelength, and repetition rate of 457 fs, 1045 nm, and 100 kHz, respectively. Two laser powers of 300 mW and 600 mW were adopted for the LAL. The spot sizes of the 300 mW LAL and 600 mW LAL were 15 and 20 μm, which gave the laser fluences were 170 of 191 J/cm^2 for the 300 mW LAL and 600 mW LAL, respectively. A circular Ag plate with a diameter of 10 mm and a thickness of 1 mm was placed inside a glass dish filled with 8 mL acetone. The liquid thickness was ca. 5 mm above the target surface. The fs laser beam was then focused on the Ag plate surface using a 20× objective lens (NA = 0.45, Mitutoyo, Japan). An area of 3.5 × 3.5 mm^2 was scanned using the parallel-line scanning method, described by the authors of [55–57], with an adjacent line interval of 5 μm. The scanning speed was set at 1 mm/s. Each ablation lasted 1 h to ensure long-period LAL [9].

The colloids that were just synthesized by the LAL, termed fresh colloids, were directly deposited onto the TEM grids (EMJapan, U1015, Tokyo, Japan, 20 nm thick carbon films on copper grids) and then characterized using TEM (Jeol, JEM-1230, Tokyo, Japan). UV-VIS spectroscopy (Shimadzu, UV-3600 Plus, Kyoto, Japan) was used to measure the absorption spectra during the colloidal aging every 30 min. More than 500 particles were measured by ImageJ to calculate the average particle sizes of the colloids. For the XRD (Rigaku, CuKα radiation (40 kV-30 mA), SmartLab-R 3kW, Tokyo, Japan) measurement, the colloids were centrifuged by a centrifuge (Eppendorf, Centrifuge 5430, Hamburg, Germany) at a rotation speed of 14,000 rpm for 10 min, and then deposited on a silicon wafer. The colloids were stored and sealed in glass containers at room temperature. After six months of storage, both the colloids in the liquids and the precipitated particles at the bottom of the glass container were characterized using TEM and UV-VIS spectroscopy. The surface chemistry of the Ag particles was analyzed using FTIR (Shimadzu, Prestige-21, Kyoto, Japan) and Raman spectrometers (LabRAM, He-Ne laser, 632 nm, 0.686 mW, Tokyo, Japan).

The nanosecond laser synthesis of the Ag colloids was performed using a Nd:YAG ns-laser (SpitLight DPSS250-100, InnoLas Laser GmbH, Krailling, Germany) at a pulse duration, wavelength, repetition rate, and pulse energy of 11 ns, 1064 nm, 100 Hz, and 150 mJ, respectively. The detected spot size on the target surface after ablation was 2.5 mm in diameter. The laser fluence is calculated to be 3 J/cm^2. The Ag target was placed inside a glass cuvette filled with 3 mL of acetone. The liquid thickness above the target corresponded to the optical path (10 mm) of the glass cuvette used. The experiments were performed under continuous stirring of the liquid, with irradiation of the unfocused laser beam for 60 s. The colloids were characterized using UV-VIS spectroscopy (Thermo Scientific Evolution 201, Tokyo, Japan) and TEM (Zeiss, Type EM 910, Oberkochen, Germany), directly after synthesis and after five weeks of storage in sealed polypropylene tubes at room temperature.

4. Conclusions

The investigation of Ag colloids synthesized by the fs-LAL of Ag in acetone, at laser powers of 300 mW and 600 mW, confirmed the spontaneous growth of the ultrasmall particles and the spontaneous size separation during long-period storage. The FTIR spectroscopy analysis showed that the CO and CO$_2$ molecules are adsorbed on the Ag particles to form Ag(CO)$_n$ and Ag(CO$_2$)$_n$ complexes, which may contribute to the high stability of the supernatant Ag NPs. Carboxylate (COO–) and hydroxyl (OH–) species also conjugate on the Ag NPs, which may be the reason for the anisotropic growth of the Ag particles into the polygonal nanocrystals over time. Both the particle size increase and the shape transformation into the polygonal nanocrystals caused a redshift of the SPR bands from 404 nm to 414 nm. The carbonaceous species generated from the acetone decomposition and pyrolysis during the LAL gradually adsorbed onto the large particles, and the soft carbon shells of the Ag@C particles also captured smaller particles, to form Ag@C-Ag aggregates. Particle aggregation and the formation of Ag–C networks significantly compromise the stability of the large Ag particles and cause their gradual precipitation during long-period storage, which leaves ultrasmall particles behind in the liquid supernatant. A higher size separation ability is endowed with the colloids obtained by fs-LAL at a higher power (600 mW), which possibly benefits from a larger amount of carbon captors, because of the stronger decomposition or pyrolysis of the solvent, which possibly creates olefinic species with a low solubility in acetone. Such phase-separating, carbon-induced size separation induced by the colloids themselves may be helpful to conquer the challenging issue of the wide (i.e., bimodal) size distribution of the metallic particles generated by ultrashort-pulsed-LAL. When the growth is terminated and mechanically fractionated from the supernatant as a precipitate, the resulting supernatant of the fs-LAL-derived colloids bears a significantly higher fraction of very small particles ($\leq$5 nm), compared to the ns-LAL. On the other hand, the ns-LAL of the Ag in acetone yields a monomodal particle size distribution with lower polydispersity, and the particle size also grows during storage, but without size focusing via precipitation, because no thick organic shells were observed by the TEM observation.

The particle growth is often neglected for the LAL-generated metallic particles. This must be taken into account when characterizing the particle state by TEM and before using these particles for practical applications. If the characterization and application were conducted at different times and particle growth had occurred, then the mismatch of the particle properties and their performances, particularly for size-sensitive applications such as catalysis and biology, would mislead the interpretation of all of the experimental results.

Author Contributions: D.Z. and K.S. conceived and designed the experiments; W.C. performed the experiments; D.Z. characterized the particles and analyzed the data; S.B., J.J., and M.-R.K. contributed with the ns-LAL work; D.Z. and K.S. wrote the paper; and all of the authors read and revised the manuscript.

Acknowledgments: We would like to thank Materials Characterization Support Unit, RIKEN CEMS for providing access to the SEM and TEM microscopes as well as XRD and FTIR spectrometers.

Conflicts of Interest: The authors declare no conflict of interest.

References

1. Zhang, D.; Gökce, B.; Barcikowski, S. Laser synthesis and processing of colloids: Fundamentals and applications. *Chem. Rev.* **2017**, *117*, 3990–4103. [CrossRef] [PubMed]
2. Zeng, H.; Du, X.W.; Singh, S.C.; Kulinich, S.A.; Yang, S.; He, J.; Cai, W. Nanomaterials via laser ablation/irradiation in liquid: A review. *Adv. Funct. Mater.* **2012**, *22*, 1333–1353. [CrossRef]
3. Xiao, J.; Liu, P.; Wang, C.X.; Yang, G.W. External field-assisted laser ablation in liquid: An efficient strategy for nanocrystal synthesis and nanostructure assembly. *Prog. Mater. Sci.* **2017**, *87*, 140–220. [CrossRef]
4. Zhang, J.; Claverie, J.; Chaker, M.; Ma, D. Colloidal metal nanoparticles prepared by laser ablation and their applications. *ChemPhysChem* **2017**, *18*, 986–1006. [CrossRef] [PubMed]
5. Zhang, J.; Chen, G.; Chaker, M.; Rosei, F.; Ma, D. Gold nanoparticle decorated ceria nanotubes with significantly high catalytic activity for the reduction of nitrophenol and mechanism study. *Appl. Catal. B* **2013**, *132*, 107–115. [CrossRef]
6. Hebié, S.; Holade, Y.; Maximova, K.; Sentis, M.; Delaporte, P.; Kokoh, K.B.; Napporn, T.W.; Kabashin, A.V. Advanced electrocatalysts on the basis of bare au nanomaterials for biofuel cell applications. *ACS Catal.* **2015**, *5*, 6489–6496. [CrossRef]
7. Zhang, D.; Liu, J.; Li, P.; Tian, Z.; Liang, C. Recent advances in surfactant-free, surface charged and defect-rich catalysts developed by laser ablation and processing in liquids. *ChemNanoMat* **2017**, *3*, 512–533. [CrossRef]
8. Zhang, J.; Chaker, M.; Ma, D. Pulsed laser ablation based synthesis of colloidal metal nanoparticles for catalytic applications. *J. Colloid Interface Sci.* **2017**, *489*, 138–149. [CrossRef] [PubMed]
9. Zhang, D.; Liu, J.; Liang, C. Perspective on how laser-ablated particles grow in liquids. *Sci. China Phys. Mech. Astron.* **2017**, *60*, 074201. [CrossRef]
10. Jendrzej, S.; Gökce, B.; Amendola, V.; Barcikowski, S. Barrierless growth of precursor-free, ultrafast laser-fragmented noble metal nanoparticles by colloidal atom clusters—A kinetic in situ study. *J. Colloid Interface Sci.* **2016**, *463*, 299–307. [CrossRef] [PubMed]
11. Poletti, A.; Fracasso, G.; Conti, G.; Pilot, R.; Amendola, V. Laser generated gold nanocorals with broadband plasmon absorption for photothermal applications. *Nanoscale* **2015**, *7*, 13702–13714. [CrossRef] [PubMed]
12. Zhou, L.; Zhang, H.; Bao, H.; Liu, G.; Li, Y.; Cai, W. Onion-structured spherical mos_2 nanoparticles induced by laser ablation in water and liquid droplets' radial solidification/oriented growth mechanism. *J. Phys. Chem. C* **2017**, *121*, 23233–23239. [CrossRef]
13. Kabashin, A.V.; Meunier, M. Synthesis of colloidal nanoparticles during femtosecond laser ablation of gold in water. *J. Appl. Phys.* **2003**, *94*, 7941–7943. [CrossRef]
14. Rehbock, C.; Merk, V.; Gamrad, L.; Streubel, R.; Barcikowski, S. Size control of laser-fabricated surfactant-free gold nanoparticles with highly diluted electrolytes and their subsequent bioconjugation. *Phys. Chem. Chem. Phys.* **2013**, *15*, 3057–3067. [CrossRef] [PubMed]
15. Liu, J.; Liang, C.; Tian, Z.; Zhang, S.; Shao, G. Spontaneous growth and chemical reduction ability of Ge nanoparticles. *Sci. Rep.* **2013**, *3*, 1741. [CrossRef]
16. Zhang, D.; Lau, M.; Lu, S.; Barcikowski, S.; Gökce, B. Germanium sub-microspheres synthesized by picosecond pulsed laser melting in liquids: Educt size effects. *Sci. Rep.* **2017**, *7*, 40355. [CrossRef] [PubMed]
17. Zhang, D.; Gökce, B.; Notthoff, C.; Barcikowski, S. Layered seed-growth of agge football-like microspheres via precursor-free picosecond laser synthesis in water. *Sci. Rep.* **2015**, *5*, 13661. [CrossRef] [PubMed]
18. Zhang, D.; Ma, Z.; Spasova, M.; Yelsukova, A.E.; Lu, S.; Farle, M.; Wiedwald, U.; Gökce, B. Formation mechanism of laser-synthesized iron-manganese alloy nanoparticles, manganese oxide nanosheets and nanofibers. *Part. Part. Syst. Charact.* **2017**, *34*, 1600225. [CrossRef]
19. Liang, C.; Sasaki, T.; Shimizu, Y.; Koshizaki, N. Pulsed-laser ablation of mg in liquids: Surfactant-directing nanoparticle assembly for magnesium hydroxide nanostructures. *Chem. Phys. Lett.* **2004**, *389*, 58–63. [CrossRef]
20. Zhang, H.; Duan, G.; Li, Y.; Xu, X.; Dai, Z.; Cai, W. Leaf-like tungsten oxide nanoplatelets induced by laser ablation in liquid and subsequent aging. *Cryst. Growth Des.* **2012**, *12*, 2646–2652. [CrossRef]
21. Niu, K.Y.; Yang, J.; Kulinich, S.A.; Sun, J.; Li, H.; Du, X.W. Morphology control of nanostructures via surface reaction of metal nanodroplets. *J. Am. Chem. Soc.* **2010**, *132*, 9814–9819. [CrossRef] [PubMed]
22. Schaumberg, C.A.; Wollgarten, M.; Rademann, K. Metallic copper colloids by reductive laser ablation of non metallic copper precursor suspensions. *J. Phys. Chem. A* **2014**, *118*, 8329–8337. [CrossRef] [PubMed]

23. Liu, J.; Liang, C.; Zhu, X.; Lin, Y.; Zhang, H.; Wu, S. Understanding the solvent molecules induced spontaneous growth of uncapped tellurium nanoparticles. *Sci. Rep.* **2016**, *6*, 32631. [CrossRef] [PubMed]

24. Wu, C.-H.; Chen, S.-Y.; Shen, P. Special grain boundaries of anatase nanocondensates by oriented attachment. *CrystEngComm* **2014**, *16*, 1459–1465. [CrossRef]

25. Wang, H.; Odawara, O.; Wada, H. Facile and chemically pure preparation of YVO_4: Eu^{3+} colloid with novel nanostructure via laser ablation in water. *Sci. Rep.* **2016**, *6*, 20507. [CrossRef] [PubMed]

26. Huang, C.-C.; Yeh, C.-S.; Ho, C.-J. Laser ablation synthesis of spindle-like gallium oxide hydroxide nanoparticles with the presence of cationic cetyltrimethylammonium bromide. *J. Phys. Chem. B* **2004**, *108*, 4940–4945. [CrossRef]

27. Xia, Y.; Xiong, Y.; Lim, B.; Skrabalak, S.E. Shape-controlled synthesis of metal nanocrystals: Simple chemistry meets complex physics? *Angew. Chem. Int. Ed.* **2009**, *48*, 60–103. [CrossRef] [PubMed]

28. Tsuji, T.; Kakita, T.; Tsuji, M. Preparation of nano-size particles of silver with femtosecond laser ablation in water. *Appl. Surf. Sci.* **2003**, *206*, 314–320. [CrossRef]

29. Pyatenko, A.; Shimokawa, K.; Yamaguchi, M.; Nishimura, O.; Suzuki, M. Synthesis of silver nanoparticles by laser ablation in pure water. *Appl. Phys. A* **2004**, *79*, 803–806. [CrossRef]

30. Streubel, R.; Bendt, G.; Gökce, B. Pilot-scale synthesis of metal nanoparticles by high-speed pulsed laser ablation in liquids. *Nanotechnology* **2016**, *27*, 205602. [CrossRef] [PubMed]

31. Tilaki, R.M.; Mahdavi, S.M. Stability, size and optical properties of silver nanoparticles prepared by laser ablation in different carrier media. *Appl. Phys. A* **2006**, *84*, 215–219. [CrossRef]

32. Tsuji, T.; Tsuji, M.; Hashimoto, S. Utilization of laser ablation in aqueous solution for observation of photoinduced shape conversion of silver nanoparticles in citrate solutions. *J. Photochem. Photobiol. A* **2011**, *221*, 224–231. [CrossRef]

33. Tsuji, T.; Mizuki, T.; Ozono, S.; Tsuji, M. Laser-induced silver nanocrystal formation in polyvinylpyrrolidone solutions. *J. Photochem. Photobiol. A* **2009**, *206*, 134–139. [CrossRef]

34. Tsuji, T.; Higuchi, T.; Tsuji, M. Laser-induced structural conversions of silver nanoparticles in pure water—Influence of laser intensity. *Chem. Lett.* **2005**, *34*, 476–477. [CrossRef]

35. Tsuji, T.; Kikuchi, M.; Kagawa, T.; Adachi, H.; Tsuji, M. Morphological changes from spherical silver nanoparticles to cubes after laser irradiation in acetone–water solutions via spontaneous atom transportation process. *Colloids Surf. A* **2017**, *529*, 33–37. [CrossRef]

36. Tsuji, T.; Nakanishi, M.; Mizuki, T.; Ozono, S.; Tsuji, M.; Tsuboi, Y. Preparation and shape-modification of silver colloids by laser ablation in liquids: A brief review. *Sci. Adv. Mater.* **2012**, *4*, 391–400. [CrossRef]

37. Kőrösi, L.; Rodio, M.; Dömötör, D.; Kovács, T.; Papp, S.; Diaspro, A.; Intartaglia, R.; Beke, S. Ultra-small, ligand-free Ag nanoparticles with high antibacterial activity prepared by pulsed laser ablation in liquid. *J. Chem.* **2016**, *2016*, 4143560. [CrossRef]

38. Tiedemann, D.; Taylor, U.; Rehbock, C.; Jakobi, J.; Klein, S.; Kues, W.A.; Barcikowski, S.; Rath, D. Reprotoxicity of gold, silver, and gold–silver alloy nanoparticles on mammalian gametes. *Analyst* **2014**, *139*, 931–942. [CrossRef] [PubMed]

39. Robertson, J. Diamond-like amorphous carbon. *Mater. Sci. Eng.* **2002**, *37*, 129–281. [CrossRef]

40. Sengan, M.; Veeramuthu, D.; Veerappan, A. Photosynthesis of silver nanoparticles using durio zibethinus aqueous extract and its application in catalytic reduction of nitroaromatics, degradation of hazardous dyes and selective colorimetric sensing of mercury ions. *Mater. Res. Bull.* **2018**, *100*, 386–393. [CrossRef]

41. Bastús, N.G.; Merkoçi, F.; Piella, J.; Puntes, V. Synthesis of highly monodisperse citrate-stabilized silver nanoparticles of up to 200 nm: Kinetic control and catalytic properties. *Chem. Mater.* **2014**, *26*, 2836–2846. [CrossRef]

42. An, J.; Tang, B.; Ning, X.; Zhou, J.; Xu, S.; Zhao, B.; Xu, W.; Corredor, C.; Lombardi, J.R. Photoinduced shape evolution: From triangular to hexagonal silver nanoplates. *J. Phys. Chem. C* **2007**, *111*, 18055–18059. [CrossRef]

43. Yu, B.; Shi, Y.; Yuan, B.; Qiu, S.; Xing, W.; Hu, W.; Song, L.; Lo, S.; Hu, Y. Enhanced thermal and flame retardant properties of flame-retardant-wrapped graphene/epoxy resin nanocomposites. *J. Mater. Chem. A* **2015**, *3*, 8034–8044. [CrossRef]

44. Mansur, H.S.; Sadahira, C.M.; Souza, A.N.; Mansur, A.A.P. Ftir spectroscopy characterization of poly (vinyl alcohol) hydrogel with different hydrolysis degree and chemically crosslinked with glutaraldehyde. *Mater. Sci. Eng. C* **2008**, *28*, 539–548. [CrossRef]

45. Yajima, T.; Uchida, H.; Watanabe, M. In-situ ATR-FTIR spectroscopic study of electro-oxidation of methanol and adsorbed CO at Pt–Ru alloy. *J. Phys. Chem. B* **2004**, *108*, 2654–2659. [CrossRef]

46. Pritchard, J.; Catterick, T.; Gupta, R.K. Infrared spectroscopy of chemisorbed carbon monoxide on copper. *Surf. Sci.* **1975**, *53*, 1–20. [CrossRef]

47. Dong, C.; Wirasaputra, A.; Luo, Q.; Liu, S.; Yuan, Y.; Zhao, J.; Fu, Y. Intrinsic flame-retardant and thermally stable epoxy endowed by a highly efficient, multifunctional curing agent. *Materials* **2016**, *9*, 1008. [CrossRef] [PubMed]

48. Liang, B.; Zhou, M.; Andrews, L. Reactions of laser-ablated Ni, Pd, and Pt atoms with carbon monoxide: Matrix infrared spectra and density functional calculations on $M(CO)_n$ (n = 1–4), $M(CO)_n^-$ (n = 1–3), and $M(CO)_n^+$ (n = 1–2), (M = Ni, Pd, Pt). *J. Phys. Chem. A* **2000**, *104*, 3905–3914. [CrossRef]

49. Ramis, G.; Busca, G.; Lorenzelli, V. Low-temperature CO_2 adsorption on metal oxides: Spectroscopic characterization of some weakly adsorbed species. *Mater. Chem. Phys.* **1991**, *29*, 425–435. [CrossRef]

50. Liang, B.; Andrews, L. Reactions of laser-ablated Ag and Au atoms with carbon monoxide: Matrix infrared spectra and density functional calculations on $Ag(CO)_n$ (n = 2, 3), $Au(CO)_n$ (n = 1, 2) and $M(CO)_n^+$ (n = 1–4; M = Ag, Au). *J. Phys. Chem. A* **2000**, *104*, 9156–9164. [CrossRef]

51. Mikhlin, Y.L.; Vorobyev, S.A.; Saikova, S.V.; Vishnyakova, E.A.; Romanchenko, A.S.; Zharkov, S.M.; Larichev, Y.V. On the nature of citrate-derived surface species on Ag nanoparticles: Insights from X-ray photoelectron spectroscopy. *Appl. Surf. Sci.* **2018**, *427*, 687–694. [CrossRef]

52. Escobar-Alarcón, L.; Espinosa-Pesqueira, M.E.; Solis-Casados, D.A.; Gonzalo, J.; Solis, J.; Martinez-Orts, M.; Haro-Poniatowski, E. Two-dimensional carbon nanostructures obtained by laser ablation in liquid: Effect of an ultrasonic field. *Appl. Phys. A* **2018**, *124*, 141. [CrossRef]

53. Van't Zand, D.D.; Nachev, P.; Rosenfeld, R.; Wagener, P.; Pich, A.; Klee, D.; Barcikowski, S. Nanocomposite fibre fabrication via in situ monomer grafting and bonding on laser-generated nanoparticles. *J. Laser Micro/Nanoeng.* **2012**, *7*, 21–27. [CrossRef]

54. Shih, C.-Y.; Streubel, R.; Heberle, J.; Letzel, A.; Shugaev, M.; Wu, C.; Schmidt, M.; Gokce, B.; Barcikowski, S.; Zhigilei, L. Two mechanisms of nanoparticle generation in picosecond laser ablation in liquids: The origin of the bimodal size distribution. *Nanoscale* **2018**, *10*, 6900–6910. [CrossRef] [PubMed]

55. Zhang, D.; Chen, F.; Fang, G.; Yang, Q.; Xie, D.; Qiao, G.; Li, W.; Si, J.; Hou, X. Wetting characteristics on hierarchical structures patterned by a femtosecond laser. *J. Micromech. Microeng.* **2010**, *20*, 075029. [CrossRef]

56. Zhang, D.; Chen, F.; Yang, Q.; Si, J.; Hou, X. Mutual wetting transition between isotropic and anisotropic on directional structures fabricated by femotosecond laser. *Soft Matter* **2011**, *7*, 8337–8342. [CrossRef]

57. Zhang, D.; Chen, F.; Yang, Q.; Yong, J.; Bian, H.; Ou, Y.; Si, J.; Meng, X.; Hou, X. A simple way to achieve pattern-dependent tunable adhesion in superhydrophobic surfaces by a femtosecond laser. *ACS Appl. Mater. Interfaces* **2012**, *4*, 4905–4912. [CrossRef] [PubMed]

nanomaterials

MDPI

Article

Magnetic Fe@FeO$_x$, Fe@C and α-Fe$_2$O$_3$ Single-Crystal Nanoblends Synthesized by Femtosecond Laser Ablation of Fe in Acetone

Dongshi Zhang [1,†], Wonsuk Choi [1,2,3,†], Yugo Oshima [4], Ulf Wiedwald [5], Sung-Hak Cho [2,6], Hsiu-Pen Lin [7,8], Yaw Kuen Li [8], Yoshihiro Ito [7,9] and Koji Sugioka [1,*]

1   RIKEN Center for Advanced Photonics, 2-1 Hirosawa, Wako, Saitama 351-0198, Japan; dongshi17@126.com (D.Z.); cws@kimm.re.kr (W.C.)
2   Department of Nano-Mechatronics, Korea University of Science and Technology (UST), 217 Gajeong-Ro, Yuseong-Gu, Daejeon 34113, Korea; shcho@kimm.re.kr
3   Department of Nano-Manufacturing Technology, Korea Institute of Machinery and Material (KIMM), 156 Gajeongbuk-Ro, Yuseong-Gu, Daejeon 34103, Korea
4   Condensed Molecular Materials Laboratory, RIKEN Cluster for Pioneering Research, 2-1 Hirosawa, Wako, Saitama 351-0198, Japan; yugo@riken.jp
5   Faculty of Physics and Center for Nanointegration Duisburg-Essen (CENIDE), University of Duisburg-Essen, 47057 Duisburg, Germany; ulf.wiedwald@uni-due.de
6   Department of Laser & Electron Beam Application, Korea Institute of Machinery and Material (KIMM), 156 Gajeongbuk-Ro, Yuseong-Gu, Daejeon 34103, Korea
7   Emergent Bioengineering Materials Research Team, RIKEN Center for Emergent Matter Science, 2-1 Hirosawa, Wako, Saitama 351-0198, Japan; hsiu-pen.lin@riken.jp (H.-P.L.); y-ito@riken.jp (Y.I.)
8   Department of Applied Chemistry, National Chiao Tung University, Science Building 2, 1001 Ta Hsueh Road, Hsinchu 300, Taiwan; ykl@cc.nctu.edu.tw
9   Nano Medical Engineering Laboratory, RIKEN Cluster for Pioneering Research, 2-1 Hirosawa, Wako, Saitama 351-0193, Japan
*   Correspondence: ksugioka@riken.jp; Tel.: +81-(0)48-467-9495
†   These authors contributed equally to this work.

Received: 24 July 2018; Accepted: 18 August 2018; Published: 20 August 2018

Abstract: There are few reports on zero-field-cooled (ZFC) magnetization measurements for Fe@FeO$_x$ or FeO$_x$ particles synthesized by laser ablation in liquids (LAL) of Fe, and the minimum blocking temperature (T_B) of 120 K reported so far is still much higher than those of their counterparts synthesized by chemical methods. In this work, the minimum blocking temperature was lowered to 52 K for 4–5 nm α-Fe$_2$O$_3$ particles synthesized by femtosecond laser ablation of Fe in acetone. The effective magnetic anisotropy energy density (K_{eff}) is calculated to be 2.7–5.4 × 10^5 J/m^3, further extending the K_{eff} values for smaller hematite particles synthesized by different methods. Large amorphous-Fe@α-Fe$_2$O$_3$ and amorphous-Fe@C particles of 10–100 nm in diameter display a soft magnetic behavior with saturation magnetization (M_s) and coercivities (H_c) values of 72.5 emu/g and 160 Oe at 5 K and 61.9 emu/g and 70 Oe at 300 K, respectively, which mainly stem from the magnetism of amorphous Fe cores. Generally, the nanoparticles obtained by LAL are either amorphous or polycrystalline, seldom in a single-crystalline state. This work also demonstrates the possibility of synthesizing single-crystalline α-Fe$_2$O$_3$ hematite crystals of several nanometers with (104), (113), (116) or (214) crystallographic orientations, which were produced simultaneously with other products including carbon encapsulated amorphous Fe (a-Fe@C) and Fe@FeO$_x$ core-shell particles by LAL in one step. Finally, the formation mechanisms for these nanomaterials are proposed and the key factors in series events of LAL are discussed.

Keywords: hematite α-Fe$_2$O$_3$; core-shell; blocking temperature; superparamagnetism; laser ablation in liquids; femtosecond laser; single-crystalline

1. Introduction

The newly emerged technique of laser ablation in liquids (LAL) [1–5] has proven to be valid for the synthesis of a large variety of colloids resulting from material removal of the substances [6]. With specific targets, simply changing the liquids for LAL allows the easy alteration of colloidal properties such as sizes [7,8] and phases [9–11]. For the synthesis of magnetic particles by LAL, the most frequently investigated material is Fe, whose results have turned out to be very diverse depending on experimental conditions [2], in which nanosecond (ns) lasers were typically used. For example, Vahabzadeh and Torkamany made use of ns lasers at fundamental and 2nd harmonic wavelengths (1064 nm and 532 nm) to ablate Fe in water for the formation of Fe_3O_4 and FeO particles, whose saturation magnetizations (M_s) and coercivity (H_c) were 22.5 emu/g and 11.5 Oe and 14.8 emu/g and 22 Oe [12], respectively. These values were much lower than those of bulk magnetite (M_s = 92 emu/g and H_c = 500–800 Oe). Zeng et al. prepared FeO nanoparticles by ns laser fragmentation in liquid (LFL) of Fe using a water/poly(vinyl pyrrolidone) (PVP) solution [13]. Amendola et al. obtained FeO_x (Fe_3O_4, FeO, α-Fe) nanoparticles with an M_s of 100 emu/g by ns LAL in water [14]. Pandey et al. found that the ns LAL of commercial Fe_2O_3 powders in doubly distilled water improved the hematite particle crystallinity and increased M_s from 0.024 to 3.41 emu/g [15]. Svetlichnyi et al. found that the H_c of the FeO_x (Fe, Fe_2O_3 and Fe_3O_4) nanoparticles synthesized by ns LAL of Fe in water increased from 144 Oe to 370 Oe when the measurement temperature was reduced from 300 K to 77 K [16]. Ismail et al. showed that the M_s values of 16.3–20.3 emu/g for the magnetic iron oxide (Fe_3O_4, α-Fe_2O_3, FeO and ε-Fe_2O_3) nanoparticles synthesized by ns LAL in SDS aqueous solution were larger than 13.8–16.2 emu/g for FeO_x (Fe_3O_4, α-Fe_2O_3, ε-Fe_2O_3) particles synthesized by ns LAL in dimethylformamide (DMF) at the same laser energies [17]. Meanwhile, Kanitz et al. employed a femtosecond (fs) laser for LAL of Fe and reported that the products changed depending on the adopted solutions: α-iron, wüstite and magnetite were synthesized in water, α-iron, cementite and FeO_x in methanol, amorphous-Fe and α-Fe mixture in ethanol and acetone, and amorphous-Fe@C core-shell particles in toluene [18]. Their M_s and H_c values were measured to be 23, 80, 60, 67 and 14 emu/g, and 77, 92, 65, 56, and 52 Oe at 300 K, respectively. Santillán et al. synthesized FeO_x (Fe_3O_4, γ-Fe_2O_3 or α-Fe) by fs laser (120 fs, 1 kHz, 800 nm) ablation of Fe in water, by which M_s, number density, magnetic radius and total radius of 49.3 emu/g, 2.9×10^{18}, 1.1 nm, 1.9 nm, respectively, were obtained, which were different from the magnetic properties (26.7 emu/g, 3468 μ_B and 0.7×10^{18}, 1.5 nm, 3.2 nm) of the particles obtained by LAL in a trisodium citrate aqueous solution [19]. Most of these works mainly focused on the hysteresis curves to reveal the dependence of magnetic properties on the phases of LAL-prepared FeO_x or Fe/FeO_x particles. Little attention has been paid to the zero-field-cooled (ZFC) and field-cooled (FC) curves including the information about blocking temperature (T_B) which is closely related to the product of particle size and magnetic anisotropy energy density essentially defining the energy barrier between two easy axes of magnetization [20].

Amendola et al. observed a blocking temperature T_B = 200 K of FeO_x particles synthesized by the ns-LAL of Fe in water [14]. Franzel et al. reported that Fe_3O_4 and Fe_3C synthesized by picosecond (ps) LAL of Fe in ethanol had a blocking temperature T_B = 120 K [21], much lower that of FeO_x particles synthesized by ns-LAL, which was due to the generation of higher ratios of smaller particles by ps-LAL. As is well known, a low T_B value (<100 K) is a good indicator for superparamagnetism arising from small particles [22,23]. The endowment of superparamagnetism to the as-prepared magnetic particles often requires the particle size to be around or less than 10 nm [24]. To date, the minimum blocking temperature of LAL-generated Fe@FeO_x or FeO_x particles is still above 100 K [21]. Due to the fact that smaller FeO_x particles often possess lower blocking temperatures [20], the synthesis of ultrasmall Fe@FeO_x and FeO_x particles by LAL is essential to lower the blocking temperature below 100 K. To this end, fs-LAL is a better choice than ps- and ns-LAL because of the phase/Coloumb explosion mechanism for fs-LAL rather than the thermal ablation mechanism for ns-LAL [25].

Despite many reports on the fs-LAL of Fe and studies on the magnetic properties (M_s, M_r and H_c) of the products [11,18], the ZFC/FC curves were not measured. To fill this gap and to further lower the blocking temperature of Fe@FeO$_x$ possessing superparamagnetic properties, the fs-LAL of Fe in acetone was performed. XRD, high-resolution transition electron microscropy (HRTEM), energy-dispersive X-ray (EDX), selected area electron diffraction (SAED), fast Fourier transform (FFT), X-ray photoelectron spectroscopy (XPS) and Raman characterizations were performed to clarify the composition of the as-prepared particles. TEM analysis was performed to display the particle morphologies and calculate the size distribution of the colloids. Both ZFC/FC and hysteresis curves of the synthesized magnetic particles were measured, from which T$_B$, M$_s$, M$_r$ and H$_c$ values were determined.

2. Materials and Methods

Colloids were synthesized by laser ablation of a Fe sheet (99.45 wt % Fe, 0.42 wt % O, 0.13 wt % C) using a fs laser system (FGPA μJewel D-1000-UG3, IMRA America Inc., Ann Arbor, MI, USA). The pulse duration, wavelength and repetition rate of the laser system were 457 fs, 1045 nm and 100 kHz, respectively. An Fe sheet with dimensions of 20 mm $\times$ 20 mm $\times$ 1 mm was placed inside a glass container and then immersed in 8 mL acetone for LAL. The liquid thickness above the target surface was kept at 5 mm. Then, a fs laser beam was focused on the Fe sheet surface by a 20$\times$ objective lens (numerical aperture (NA) = 0.4, Mitutoyo, Kawasaki, Japan) and scanned over an area of 3.5 $\times$ 3.5 mm^2 using the scan method described in [26–28] with a line interval of 5 μm and a scan speed of 1 mm/s to ablate the Fe sheet. The ablation process lasted around 1 h. The average laser power was set to 600 mW. The spot size was 26 μm. The peak irradiance and laser fluence were calculated to be 1.13 $\times$ 10^9 W/m^2 and 113 J/cm^2, respectively.

The colloids were directly deposited onto TEM grids (EMJapan, U1015, Tokyo, Japan, 20 nm thick carbon films on copper grids) after LAL without any pre-treatment and then characterized using TEM (Jeol, JEM-1230, Tokyo, Japan) operating at 80 kV. HRTEM and STEM-EELS (scanning transmission electron microscopy–electron energy loss spectroscopy) were performed with a JEM-ARM200F (Jeol, Tokyo, Japan) equipped with third-order aberration correctors for both illuminating and imaging lens systems operated at 200 kV. For XRD and magnetic property measurements, the colloids with liquids were centrifuged by a centrifuge (Eppendorf, Centrifuge 5430, Hamburg, Germany) at a rotation speed of 14,000 rpm for 10 min. The precipitated particles were then collected in a cuvette and dried in a freeze dryer (Rikakikai, S-1000, Eyela, Tokyo, Japan). The dried particles were deposited on an amorphous glass plate (10 mm $\times$ 10 mm $\times$ 1 mm) for XRD, XPS and Raman characterizations. The composition of the particles was analysed using XRD (Rigaku, CuKα radiation (40 kV-30 mA), SmartLab-R 3kW, Tokyo, Japan). The surface chemistry of the particles was analysed by XPS (Thermo Scientific, ESCALAB 250, Tokyo, Japan) and Raman spectroscopy (LabRAM, Hiriba, He-Ne laser, 632 nm, 0.686 mW, Tokyo, Japan). A zeta-potential and particle size analyzer (ELSZ-2PL, Photal, Osaka, Japan) was used to measure the zeta potential of the fresh colloid and the colloid stored after 3 weeks. UV-vis spectroscopy (Shimadzu, UV-3600 Plus, Tokyo, Japan) was used to measure the absorption spectra of colloids.

Magnetic properties of the particles were measured in He gas atmosphere using the superconducting quantum interference device (SQUID) magnetometer (Quantum Design, MPMS XL7, San Diego, USA). The dried 26.5 mg Fe@α-Fe$_2$O$_3$ particle powder was filled into a capsule, which was then placed into the magnetometer. Zero-field-cooled (ZFC) magnetization was measured by cooling samples in a zero magnetic field and then increasing the temperature from 5 K to 300 K with magnetization-temperature data recorded every 5 K at an applied field of 50 Oe. Field-cooled (FC) curves were recorded by cooling the samples from 300 K to 5 K with a constant field of 50 Oe. The field dependence of the magnetization (hysteresis loop) was recorded up to $\pm$70 kOe at T = 5 K and $\pm$10 kOe at T = 300 K, respectively.

3. Results

3.1. Material Property

After drying on TEM grids, the particles synthesized in acetone form a particle network (Figure 1a–c), connected by a large amount of small clusters, which is a typical phenomenon after ferro-fluidic colloid drying [29–34]. The average size of the particles is estimated to be 5–6 nm (Figure 1d). Small particles with sizes of less than 10 nm occupy more than an 87% number frequency of all of the particles. In particular, small particles with a ~90% number frequency of 1–10 nm are in the majority, with the highest number frequency at 4–5 nm (Figure 1f–i). The large particles are in the form of core-shell particles with a shell thickness of ca. 6 nm (Figure 1c). No crystalline peak was observed in the XRD spectrum (Figure 1e), similar to the case of the nanomaterials obtained by LAL of Cu in acetone [35]. This is either due to the low amount of particle powders used for the XRD spectrum or due to the too-small crystallites of the particles [35]. Actually, we used several mg particles for XRD characterization. With the same amount, a well featured XRD spectrum of Ag particles synthesized by LAL of Ag in acetone has been observed [36], which indicates that the particles synthesized by LAL of Fe in acetone are in very low crystallinity.

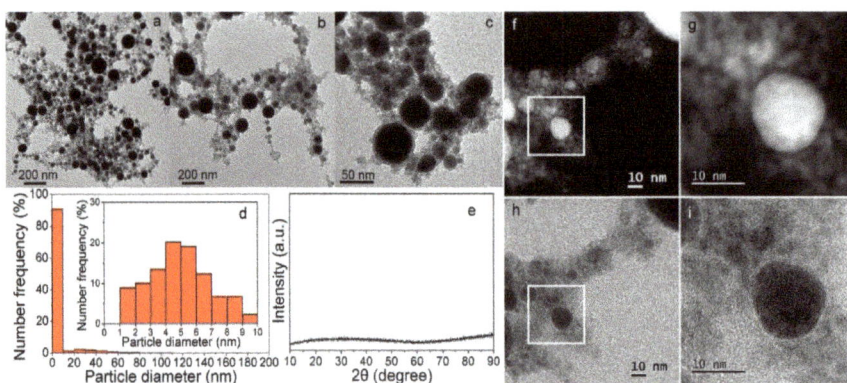

Figure 1. (a–c) TEM images of the particles synthesized by the laser ablation of Fe in acetone at 400 mW. (d) Size distribution of the synthesized particles. The inset figure shows the detailed size distribution in the range of 0–10 nm. (e) XRD pattern of the synthesized particles, where no peaks were detected, probably due to the low crystallinity of the particles and a large amount of carbon clusters. (f,g) and (h,i) black field and white field scanning transmission electron microscopy (STEM) images of small particles, respectively.

To confirm the compositions of small clusters and core-shell particles, the distribution of Fe, C, O elements in the nanoblends was analyzed by EDX as shown in Figures 2 and 3. As indicated by TEM image (Figure 2a) and the Fe and O elemental distributions (Figure 2c,e), the small clusters were identified as FeO_x. Besides this, a certain amount of carbon was also detected (Figure 2b,d). However, given that the particles were deposited on the carbon membrane of a TEM grid, it is difficult to differentiate whether the detected carbon comes from the particles or not.

Apparent evidence for the generation of carbon during LAL was witnessed by HRTEM characterization (Figure 3a,f,g) and EDX analysis (Figure 3b–e). Both amorphous carbon (Figure 3b,d,f) and onion-like carbon (Figure 3b,d,f,g) were discovered, which encapsulated amorphous Fe particles to form the amorphous-Fe@carbon (a-Fe@C) core-shell particles (a representative particle is shown with an arrow marked in Figure 3b). Facilitated by the adhesion of different carbon shells, a-Fe@C core-shell particles gradually evolve into a particle network (Figures 1a–c and 3g). This phenomenon is consistent with previous reports that LAL in organic solvents often causes the decomposition of solvent molecules and results in the formation of carbon-encapsulated particle networks [10,37]. Amorphous FeO_x clusters (Figure 3a–e) were also generated, which surrounded big core-shell particles to facilitate the formation of particle network. Figures 2a–e and 3a–e also indicate that, besides a-Fe@C core-shell particles, large $Fe@FeO_x$ core-shell particles with diameters of tens of nm are produced by LAL of Fe in acetone.

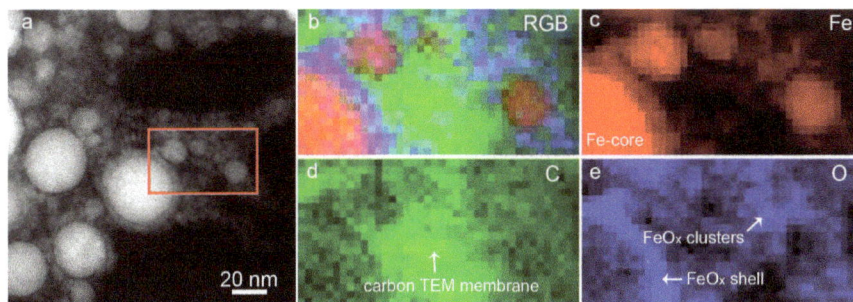

Figure 2. (a) TEM image and (b–e) EDX mapping of small clusters. (b) TEM image of mixed elements of (c) Fe, (d) C and (e) O images.

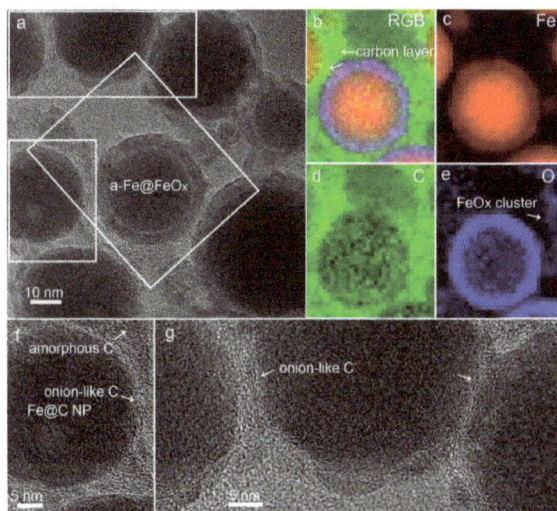

Figure 3. (a) TEM image and (b–e) EDX mapping of a $Fe@FeO_x$ core-shell particle. (b) TEM image of mixed elements of (c) Fe, (d) C and (e) O images. (f,g) TEM images of the Fe@C particles.

To better understand the compositions of $Fe@FeO_x$ core-shell particles, a core-shell particle was selected as the representative particle for HRTEM, SAED and FFT characterizations, as shown in Figure 4a–c. To further clarify the crystallinity in different regions, HRTEM images of seven domains of FeO_x shells and one larger domain of an Fe core were displayed in Figure 4d–k. The SAED pattern of the core-shell particle indicates that the core-shell particle has a low crystallinity since only two diffraction rings were observed (Figure 4b), which fits well with the (104) and (214) planes of α-Fe_2O_3 (ICSD No. 01-089-0597). FFT analysis gives more information about the crystallinity of the core-shell particle, which indicates that two more planes of the (113) and (116) planes of α-Fe_2O_3 are also present. Figure 4d–h display the HRTEM images of different crystal domains in the FeO_x shell which possess

(104), (113), (116) and (214) planes of α-Fe_2O_3 with interplanar distances of 0.270, 0.217, 0.170 and 0.150 nm, respectively. Besides α-Fe_2O_3, another diffraction ring belonging to a crystal plane with an interplanar distance of 0.300 nm was also detected, which can be assigned to the (220) plane of Fe_3O_4 (ICSD No. 01-089-0950). It is noteworthy that (i) the crystals in the FeO_x shell are mainly single crystalline; and (ii) the quality of the single-crystallinity is not particularly good because many defects among the crystal planes are obvious (Figure 4d–g,l–o). One domain of the FeO_x shell is almost completely amorphous (Figure 4j). As concluded from Figure 4d–h, the FeO_x shell is composed of many α-Fe_2O_3 single-crystalline crystals with different crystallographic orientations. Regarding the Fe core, it is totally amorphous (Figure 4k). That is why no peak was detected during XRD characterization (Figure 1e). Regarding the dominant small particles, four crystalline domains outside the $Fe@FeO_x$ particle are shown in Figure 4l–o. The interplanar distances of 0.217, 0.217, 0.217 and 0.270 nm of these crystalline domains are well indexed to (113), (113), (113) and (104) planes of α-Fe_2O_3. As indicated by HRTEM analysis, it is clear that single-crystalline α-Fe_2O_3 nanocrystals with different crystallographic orientations are abundant in the nanoblends.

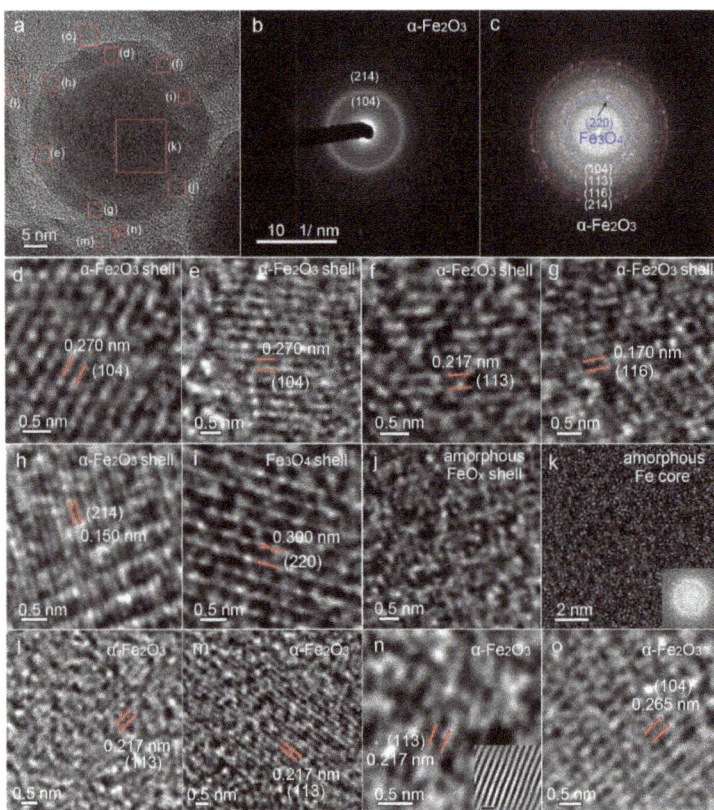

Figure 4. (**a**) High-resolution transition electron microscropy (HRTEM) image of a $Fe@FeO_x$ core-shell particle, (**b**) selected area electron diffraction (SAED) and (**c**) FFT analysis of the core-shell particle. HRTEM images of (**d–j**) α-Fe_2O_3 shells, (**i**) Fe_3O_4 shell and (**j**) amorphous FeO_x shell and (**k**) amorphous Fe core, respectively. (**l–o**) Single-crystalline α-Fe_2O_3 crystals. Inset images in (**k,n**) show the FFT image indicating an amorphous structure and the inverse FFT image of the structure showing clearer crystal planes, respectively. The scale bar in the inset image of (**n**) is 0.5 nm.

Raman spectroscopy shows seven peaks at 220.9, 239.5, 286.0, 401.4, 495.9, 606.7, 659.4, 812.0, 1050, 1099.4, 1297.7 and 1603.0 cm^{-1} (Figure 5) associated with α-Fe$_2$O$_3$, in accordance with the conclusion from HRTEM analysis (Figure 4) that α-Fe$_2$O$_3$ is the main crystalline product. Besides the peaks corresponding to α-Fe$_2$O$_3$, another small peak is also observed at 1584.7 cm^{-1}, which can be assigned to the G-band of carbon and therefore indicates the presence of a large amount of carbon in the particles, in accordance with the HRTEM images shown in Figure 3f–g and XPS analysis which show that the outermost 5 nm-thick surfaces of the as-prepared particles are composed of 72.22% C, 1.29% Fe and 26.49% O. The high C ratio and low Fe and O ratios suggest that a higher ratio of carbon/carbon-byproduct clusters are generated by the laser-induced decomposition of acetone molecues. High-resolution XPS Fe 2p, O 1s and C 1s spectra are shown in Figure 6. After peak fitting, the ratio of sp^2/sp^3-C was calculated to be 0.89. The sp^2 and sp^3 carbons correspond to an ordered graphite (sp^2) structure and disordered graphite layers (e.g., soot, chars, glassy carbon, and evaporated amorphous carbon [36,38]), respectively. Thus, more than half of the C that precipitates on Fe@Fe$_2$O$_3$ particles is crystalline because the sp^2/sp^3 ratio is less than 1, in accordance with the HRTEM images shown in Figure 3f–g, where onion-like carbons with some defects appear as shells to embed a-Fe particles inside. When both sp^2 and sp^3 C states are mixed in particles, diamond-like carbon (DLC) [39] structures are considered to be generated. DLC structures should be a typical product of LAL in organic solvents since they were also observed from other nanomaterials obtained by the LAL of different metals (e.g., Ag [36], Mo [39], Ti, [40], Ta [41], Nb [41], Hf [41], Mo [41] and Co [37]) in organic solvents. Peak fitting of C 1s (Figure 6c) shows that 15.91% and 21.90% of the carbon have C=O and C–O bonding, respectively, which is due to adsorbed acetone molecules and their decomposition byproducts. From the XPS Fe 2p spectrum (Figure 6a), only the binding energies that correspond to Fe^{3+} (711.2 eV and 724.9 eV) and Fe^0 (707.3 eV and 720.1 eV) are observed, which come from α-Fe$_2$O$_3$ and a-Fe, respectively. Because of abundant C–O and C=O bindings on the particle surfaces, only a small amount of Fe-O binding can be deconvoluted from the O 1s spectrum (Figure 6b). Both XPS and Raman spectra support the conclusion from HRTEM analysis that α-Fe$_2$O$_3$ is the dominant phase of the crystalline particles.

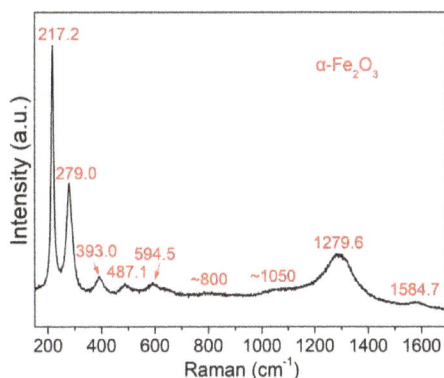

Figure 5. Raman spectrum of the particles obtained by the laser ablation in liquids (LAL) of Fe in acetone.

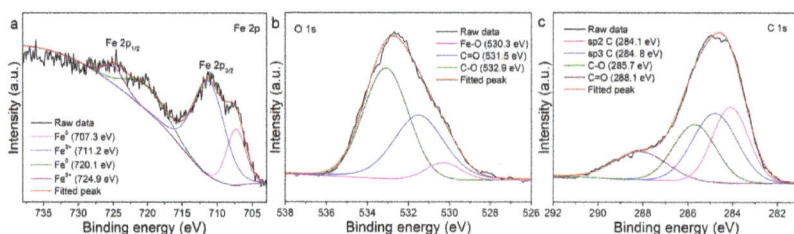

Figure 6. High resolution XPS (**a**) Fe 2p, (**b**) O 1s and (**c**) C 1s spectra from the colloids synthesized by laser ablation of Fe in acetone.

The synthesized Fe@α-Fe$_2$O$_3$ and α-Fe$_2$O$_3$ particles were unstable with gradual particle precipitation at the bottom of the glass container (right optical image inset in Figure 7a) during the colloid storage. As a result, the absorbance spectrum of the colloid downshifted (Figure 7a) and the optical transparency of the colloid increased after 3-week storage (left and middle optical images inset in Figure 7a). Additionally, the zeta potential values of the colloid decreased from −34.88 mV to −27.95 mV after 3-week storage, which indicates the decreased stability of the colloid. The zeta potential is indicative of the difference in the electric potentials between the charges of the species which strongly adsorb on the particle surface and those (with the opposite sign) of the diffuse layer in the dispersing medium [42]. It is often considered that the colloids with a zeta potential value smaller than −30 mV or larger than 30 mV are stable, while those with zeta potential values in the range of −30~30 mV are unstable [3]. Hence, in principle, one would expect that the stable particles with larger charges remain well dispersed while those with lower charges precipitate. However, in our case, the zeta potential value of the colloids decreased over time, which indicated that the gradual aggregation of nanoparticles occurred during colloidal storage. The magnetic properties among magnetic particles and the "capture" behavior of both carbon shells (Figure 3a) and free carbon clusters [36] cause colloidal aggregation and precipitation during storage. Considering the excellent long-term (six-month) stability of Ag colloid produced by LAL in acetone [36], it is highly possible that the magnetostatic interaction among magnetic particles is the main reason to be responsible for the colloidal aggregation and precipitation.

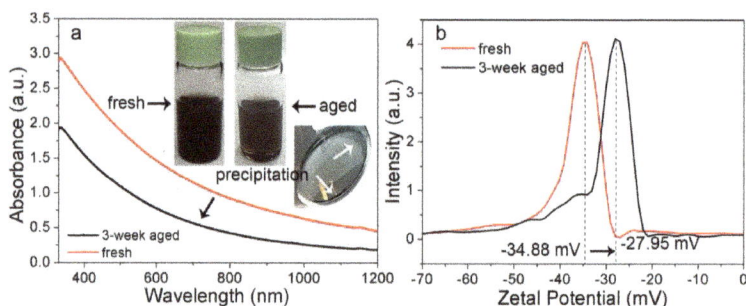

Figure 7. (**a**) Absorption spectra for the fresh colloid (red curve) synthesized by laser ablation of Fe in acetone at 600 mW and (**b**) the colloid stored for 3 weeks (black curve), respectively. Inset images in (**a**) are optical images of the fresh colloid (left) and the colloid stored for 3 weeks (middle), where the precipitation of the colloid (as indicated by white arrows in the right optical image) causes the downshift of the absorbance spectra. (**b**) Zeta potential curves of the fresh colloid and the colloid stored for 3 weeks.

Compared with the techniques of laser target evaporation in gases and laser ablation in air whereby metal-oxide [43,44] is produced, LAL is better at the synthesis of Fe@C and Fe@FeO$_x$ core-shell particles and further indicates a new way to synthesize single-crystalline iron oxide particles.

3.2. Magnetic Properties

The magnetic properties of the Fe@α-Fe$_2$O$_3$ particles prepared in acetone are presented in Figure 8. To reveal the relationship between magnetization and temperature, ZFC and FC curves of the as-prepared nanomaterials were measured. The ZFC curve in Figure 8a shows a broad maximum peaking at 52 K, followed by a plateau and a subsequent gradual increase from 250–300 K suggesting two different fractions of particles. The FC branch exhibits an almost linear increase of the magnetic moment with decreasing temperature from 300 K to 5 K. Figure 8b,c show the magnetic hysteresis loops at 5 K and 300 K with saturation magnetization M_S = 72.5 emu/g and 61.9 emu/g, respectively. The magnification around zero field delivers coercivities of H_C = 160 Oe at 5 K and 70 Oe at 300 K.

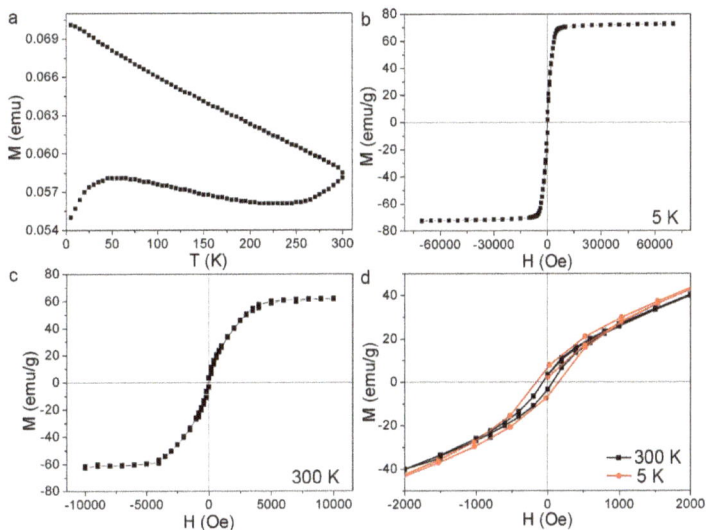

Figure 8. Magnetic characterization of the Fe@α-Fe$_2$O$_3$ particles synthesized by laser ablation in acetone. (**a**) ZFC/FC curves in 50 Oe, (**b**,**c**) hysteresis curves at 5 and 300 K. (**d**) Magnified hysteresis curves at 5 and 300 K.

The structural and morphological studies above reveal a mixture of Fe@α-Fe$_2$O$_3$, Fe@C, and α-Fe$_2$O$_3$ particles (Figures 2–4) which can be split into two size regimes. The sizes of large Fe@α-Fe$_2$O$_3$ and Fe@C core-shell particles with a number frequency of about 10% are in the range of 10–100 nm, with the maximum diameter of the distribution being 30 nm (Figure 1d). α-Fe$_2$O$_3$ particles are significantly smaller, with a medium size of 4–5 nm (Figure 1d), at a number frequency of about 90%. For the magnetometry of magnetic powders, magnetization is the ratio of the magnetic moment with respect to mass of particles in different size regimes. Taking identical mass densities as a rough estimate, the relative mass fraction of the large particles is calculated to be 87%–93%. Further consideration of the significantly smaller magnetization of α-Fe$_2$O$_3$ of 1 emu/g as compared to amorphous Fe with M_S > 100 emu/g [45] further reduces the relative magnetic signal from α-Fe$_2$O$_3$. This means that the vast majority of the magnetic signal stems from the larger particles while the smaller are expected contributing less than 1% in the saturated state.

Therefore, the saturation magnetization of M_S = 72.5 emu/g and 61.9 emu/g at 5 K and 300 K, respectively, is considered to mainly stem from the amorphous Fe cores of large a-Fe@FeO$_x$ core-shell particles. This deduction is also helpful in understanding the magnetic properties (M_S = 67 emu/g at 300 K) of a-Fe@C and Fe$_3$O$_4$ nanoblends synthesized by fs-LAL of Fe in acetone under other conditions (35 fs, 800 nm, 5 kHz, 800 µJ/pulse) [18]. The remanence to saturation ratio (M_r/M_s), also called

115

the saturation magnetization ratio, was calculated to be ca. 0.1, which was smaller than that (0.5) expected theoretically for randomly oriented single domain grains [46], which indicates the presence of a significant amount of superparamagnetic small particles, domain walls in large particles, and the occurrence of antiferromagnetic interactions [47]. Here, all the above may add to an overall low remanence. It is well known that α-Fe$_2$O$_3$ is weakly ferromagnetic or antiferromagnetic [48], and its presence as the shell material endows antiferromagnetic properties to the Fe@α-Fe$_2$O$_3$ particles while the small α-Fe$_2$O$_3$ particles are expected to be superparamagnetic at 300 K and thermally blocked at 5 K (see discussion below). The interfacial magnetic interactions between ferromagnetic cores and antiferromagnetic shells [49,50] may also endow high orbital magnetic moment to LAL-synthesized a-Fe@α-Fe$_2$O$_3$ particles. For the metallic a-Fe core, Grinstaff et al. have confirmed that glassy a-Fe is a soft ferromagnetic material with M_S = 152 emu/g and H_c = 160 Oe at T = 5 K [45]. While H_c fits well to the present results, the lower M_S can be explained by the mixture of amorphous Fe, the weakly ferromagnetic α-Fe$_2$O$_3$, and the unknown but significant amount of C in the sample. Thus, the remanent magnetization mainly originates from large Fe@α-Fe$_2$O$_3$ core-shell ferromagnetic particles with multi-domains, which cannot rapidly demagnetize by domain formation in the absence of an applied field.

A more interesting phenomenon is the magnetic signature of small α-Fe$_2$O$_3$ particles, which results in different ZFC/FC curves as compared to those of a-Fe@C particles synthesized by fs-LAL of Fe in acetone under other conditions [18]. The broad peak with a maximum at 52 K is ascribed to the blocking behavior of α-Fe$_2$O$_3$ particles on top of an almost constant signal in the interval of 5–250 K arising from the larger Fe@α-Fe$_2$O$_3$ particles. The smallest particles of the Fe@α-Fe$_2$O$_3$ particles gradually cross their blocking temperature T_B with the temperature increasing above 250 K. At 300 K, however, only a minor fraction of Fe@α-Fe$_2$O$_3$ is superparamagnetic, explaining the monotonous increase in the FC branch; moreover, the interparticle interactions [51] among LAL-generated particles [43] of the magnetic core-shell particles are supposedly not strong enough to cause collective magnetic freezing to enter a spin-glass state.

We do not observe any sharp change of the magnetization, which indicates the absence of the Morin transition in α-Fe$_2$O$_3$. Previous reports have shown that in small α-Fe$_2$O$_3$ particles with diameters below 19 nm (cf. Figure 1a–d), the Morin transition occurring in bulk α-Fe$_2$O$_3$ is smeared out over a wide temperature range or even completely suppressed [52,53]. This leads to weakly ferromagnetic α-Fe$_2$O$_3$ for all considered temperatures. The blocking behavior with T_B = 52 K at the peak position can be translated to an effective magnetic anisotropy energy density K_{eff} via $K_{eff} \cdot V$ = 25 $k_B T_B$ with V the particle volume and k_B Boltzmann's constant. The factor of 25 is the natural logarithm of the product of the measurement time window of SQUID magnetometry (10 s) and the intrinsic attempt frequency of about 10^{10} Hz [53]. For 4–5 nm hematite nanospheres, as in our cases, K_{eff} is calculated to be 2.7–5.4 $\times$ 10^5 J/m^3. Bödker et al. extracted an energy barrier of 300–600 K for 16 nm which translates to K_{eff} = 0.5–1.1 $\times$ 10^5 J/m^3 [53] and a strongly increasing K_{eff} when the particle size is reduced, reaching a maximum value of 2.4 $\times$ 10^5 J/m^3 for 5.9 nm particles at the smallest investigated diameter [54]. In this light, the obtained results for the 4–5 nm α-Fe$_2$O$_3$ particles convincingly extend the size dependence to smaller diameters.

Despite a very broad size distribution, the α-Fe$_2$O$_3$ particles synthesized by fs-LAL possess the lowest blocking temperature of 52 K among all α-Fe$_2$O$_3$ particles synthesized by LAL [2]. The blocking temperature of FeO$_x$ particles synthesized by the ns-LAL of Fe in water was ca. 220 K, which corresponds to the particle size of 15 nm [14]. The ps-LAL of Fe in ethanol gave rise to the formation of Fe$_3$O$_4$/Fe$_3$C mixture colloids which had a bimodal size distribution with maxima at ca. 3 nm and ca. 12 nm [21]. Because of the generation of a greater amount of small particles by ps-LAL, the blocking temperature down-shifted to 120 K when the applied field was 50 Oe [21]. According to the previously reported relationship between the blocking temperatures and particle sizes [20], it is estimated that the average size of Fe$_3$O$_4$/Fe$_3$C mixture colloid obtained by ps-LAL in ethanol was ca. 9 nm. In the case of fs-LAL shown in this work, the blocking temperature is further lowered to 52 K, which corresponds to the particle size of ca. 4–5 nm for α-Fe$_2$O$_3$.

3.3. Formation Mechanism

Considering the advantage of fs-LAL over ps-LAL, and ns-LAL enabling the synthesis of ultrasmall magnetic particles with lower blocking temperatures, the formation mechanism of both Fe@C, Fe@FeO$_x$ and ultrasmall α-Fe$_2$O$_3$ particles is here proposed to show the uniqueness of the fs-LAL process. The large size difference between ultrasmall α-Fe$_2$O$_3$ clusters of several nm and large core-shell particles with sizes ranging from tens of nm to 130 nm indicates that large particles do not form through particle growth mechanism but form through ejection of large Fe particles during LAL [55]. The ultrasmall α-Fe$_2$O$_3$ particles less than 10 nm (Figure 5g) should form due to phase/Coloumb explosion mechanism for fs-LAL. In contrast, owing to the thermal ablation mechanism, the sizes of the majority of the small particles inside the cavitation bubble are already 12 nm for ns-LAL [56]. The particle growth after bubble collapse often leads to a further increase in the particle sizes. Therefore, despite a small amount of large Fe particle ejection due to thermal effects during fs-LAL, the main "cold" process of fs-LAL is more efficient at generating ultrasmall particles than both ps- and ns-LAL.

Due to the plasma-induced decomposition of acetone molecules and the dissociation of the dissolved oxygen (Figure 9a) [57], O radicals are generated during fs-LAL of Fe in acetone, which may react with the surrounding Fe atoms (generated from plasma-induced target material atomization) to form FeO$_x$ clusters (Figure 9b). However, due to the existence of limited oxygen, the main products generated from the plasma phase are pure Fe clusters. The sizes of Fe and FeO$_x$ clusters may increase slightly during bubble expansion (Figure 9c) by coalescence. Inside the cavitation bubble, reductive gases [58] such as H$_2$, CO and CH$_4$ reduce FeO$_x$, which results in the formation of pure Fe particles (Figure 8d). After bubble collapses (Figure 8e), the outer parts of large Fe particles are oxidized into FeO$_x$ shells containing a large amount of α-Fe$_2$O$_3$ domains (Figure 9g), while the ultrasmall Fe particles are oxidized into α-Fe$_2$O$_3$ particles (Figure 9f,g). Due to the difference in local temperature and pressure as well as the variation in oxygen abundance around Fe particles, Fe particles crystallize into α-Fe$_2$O$_3$ or Fe$_3$O$_4$ single crystals along different crystallographic orientations. It is also possible that (1) reductive gases in the cavitation bubbles inhibit complete oxidation of Fe into FeO$_x$ and their further polycrystallization; (2) during bubble collapses, the shock waves [59] render small α-Fe$_2$O$_3$ crystals with high kinetic energy to make them quickly eject towards the already formed Fe@FeO$_x$ particles to be captured by FeO$_x$ shells as single domains.

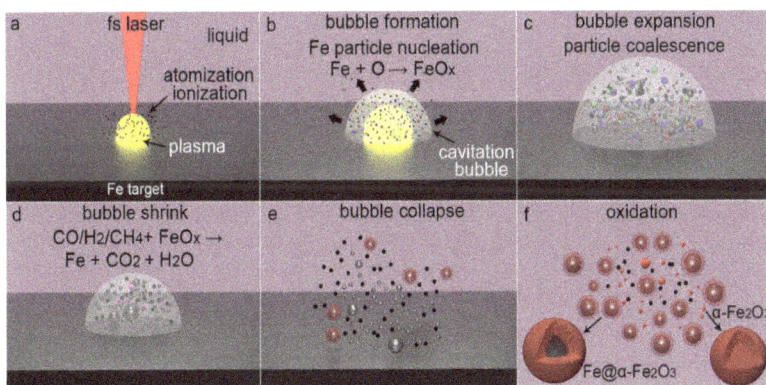

Figure 9. (a–f) Schematic of formation mechanism for Fe@α-Fe$_2$O$_3$ particles by fs-LAL of Fe in acetone. Oxygen radicals that react with Fe atoms to from FeO$_x$ come from fs laser induced decomposition of acetone and dissolved oxygen. Fe: grey color, C: black color, α-Fe$_2$O$_3$: red color, other FeO$_x$ phases: green and blue color. Note that the shockwaves [59] generated during bubble collapse can push the already formed α-Fe$_2$O$_3$ ultrasmall particles towards Fe@FeO$_x$ particles to be captured by the FeO$_x$ shells.

Extremely superfast cooling of the molten Fe droplets inhibits their crystallization so that a-Fe rather than crystalline Fe particles form after LAL. The cooling rate required for a-Fe formation ranges from 10^5 to 10^7 K/s [60], which can be easily obtained during the ultrafast quenching of LAL-generated plasma (thousands of Kelvin quenching within a submicro-second interval [61]). During the ejection of the molten Fe particles from the ablated target, they interact with acetone molecules, as a case of the electric explosion of steel in carbon-rich liquids [62,63], resulting in the formation of carbon atoms and other carbonaceous byproducts, which then precipitate on Fe particles to form C-shells. During their precipitation, Fe particles with high surface activity act as catalysts to facilitate the formation of onion-like carbon shells. The presence of carbons on the Fe particles inhibits particle growth and coalescence [4] and prevent surface oxidation, leading to the formation of Fe@C core-shell particles (Figure 3f–g).

Under the impact of the high-temperature and high-pressure environment in the plasma phase of fs-LAL, the atomization/ionization of carbon impurities in the iron substrates and the decomposition of the acetone molecules [64,65] occur simultaneously (Figure 9a), which leads to the formation of free C clusters. Besides pure carbon, polycyclic structures may be also generated through the interaction between Fe particles and organic solvents [63], which may precipitate on the already particles to make them evolve into networks. In consequence, the stability of the colloids decreases during storage and results in the precipitation of particles (Figure 7).

4. Conclusions

This work has demonstrated the capability of synthesizing large Fe@α-Fe$_2$O$_3$ and small α-Fe$_2$O$_3$ particles which are practically completely split in the size histogram. Four to five nanometer α-Fe$_2$O$_3$ particles exhibit a low blocking temperature of 52 K by fs-LAL in acetone, among the lowest ever achieved by LAL. From superparamagnetic blocking, an effective magnetic anisotropy $K_{eff} = 2.7$–5.4×10^5 J/m^3 has been estimated which extends previous investigations convincingly towards smaller hematite particle sizes. Surprisingly, most small α-Fe$_2$O$_3$ particles were single-crystalline, and so the possibility of synthesizing single-crystalline particles by LAL was demonstrated. Because of the dominant mass of large Fe@α-Fe$_2$O$_3$ and Fe@C particles (10–100 nm), all nanoblends show a soft magnetic behavior with saturation magnetization (M$_s$) and coercivities (H$_c$) values of 72.5 emu/g and 160 Oe at 5 K and 61.9 emu/g and 70 Oe at 300 K, respectively, which mainly originate from amorphous Fe core particles. Previously, ZFC/FC curves were seldom investigated as compared to hysteresis curves for LAL-generated magnetic particles. Here, it was shown that the blocking temperatures in the ZFC curves can be used to estimate the sizes of small magnetic particles.

Author Contributions: D.Z., K.S. conceived and designed the experiments; W.C. performed the experiments; D.Z. characterized the particles and analyzed the data; Y.O. and D.Z. did the magnetic measurement; U.W. analyzed the magnetic properties; H.-P.L. and Y.I. measured the zeta potential of the colloid, D.Z. and K.S. wrote the paper, all authors read and revised the manuscript.

Acknowledgments: The authors acknowledge Aiko Nakao from the Institute of Physical and Chemical Research (RIKEN) for her help with XPS measurement and analysis. Also, we would also like to thank the Materials Characterization Support Unit, RIKEN CEMS for providing access to the TEM microscopy, XRD, Raman and XPS instruments and the support from Masaki Takeguchi from NIMS (National Institute for Materials Science) for HRTEM characterization and analysis.

Conflicts of Interest: The authors declare no conflict of interest.

References

1. Zhang, D.; Gökce, B. Perspective of laser-prototyping nanoparticle-polymer composites. *Appl. Surf. Sci.* **2017**, *392*, 991–1003. [CrossRef]
2. Zhang, D.; Gökce, B.; Barcikowski, S. Laser synthesis and processing of colloids: Fundamentals and applications. *Chem. Rev.* **2017**, *117*, 3990–4103. [CrossRef] [PubMed]
3. Zhang, D.; Liu, J.; Li, P.; Tian, Z.; Liang, C. Recent advances in surfactant-free, surface charged and defect-rich catalysts developed by laser ablation and processing in liquids. *ChemNanoMat* **2017**, *3*, 512–533. [CrossRef]

4. Zhang, D.; Liu, J.; Liang, C. Perspective on how laser-ablated particles grow in liquids. *Sci. China Phys. Mech. Astron.* **2017**, *60*, 074201. [CrossRef]

5. Amendola, V.; Meneghetti, M. What controls the composition and the structure of nanomaterials generated by laser ablation in liquid solution? *Phys. Chem. Chem. Phys.* **2013**, *15*, 3027–3046. [CrossRef] [PubMed]

6. Zhang, D.; Gökce, B.; Sommer, S.; Streubel, R.; Barcikowski, S. Debris-free rear-side picosecond laser ablation of thin germanium wafers in water with ethanol. *Appl. Surf. Sci.* **2016**, *367*, 222–230. [CrossRef]

7. Letzel, A.; Gökce, B.; Wagener, P.; Ibrahimkutty, S.; Menzel, A.; Plech, A.; Barcikowski, S. Size quenching during laser synthesis of colloids happens already in the vapor phase of the cavitation bubble. *J. Phys. Chem. C* **2017**, *121*, 5356–5365. [CrossRef]

8. Rehbock, C.; Merk, V.; Gamrad, L.; Streubel, R.; Barcikowski, S. Size control of laser-fabricated surfactant-free gold nanoparticles with highly diluted electrolytes and their subsequent bioconjugation. *Phys. Chem. Chem. Phys.* **2013**, *15*, 3057–3067. [CrossRef] [PubMed]

9. Zhang, D.; Ma, Z.; Spasova, M.; Yelsukova, A.E.; Lu, S.; Farle, M.; Wiedwald, U.; Gökce, B. Formation mechanism of laser-synthesized iron-manganese alloy nanoparticles, manganese oxide nanosheets and nanofibers. *Part. Part. Syst. Charact.* **2017**, *34*, 1600225. [CrossRef]

10. Amendola, V.; Riello, P.; Meneghetti, M. Magnetic nanoparticles of iron carbide, iron oxide, iron@iron oxide, and metal iron synthesized by laser ablation in organic solvents. *J. Phys. Chem. C* **2011**, *115*, 5140–5146. [CrossRef]

11. Lasemi, N.; Bomati Miguel, O.; Lahoz, R.; Lennikov, V.; Pacher, U.; Rentenberger, C.; Kautek, W. Laser-assisted synthesis of colloidal fe$_x$o$_y$ and fe/fe$_x$o$_y$ nanoparticles in water and ethanol. *Chemphyschem* **2018**. [CrossRef] [PubMed]

12. Vahabzadeh, E.; Torkamany, M.J. Iron oxide nanocrystals synthesis by laser ablation in water: Effect of laser wavelength. *J. Clust. Sci.* **2014**, *25*, 959–968. [CrossRef]

13. Liu, P.; Cai, W.; Zeng, H. Fabrication and size-dependent optical properties of feo nanoparticles induced by laser ablation in a liquid medium. *J. Phys. Chem. C* **2008**, *112*, 3261–3266. [CrossRef]

14. Amendola, V.; Riello, P.; Polizzi, S.; Fiameni, S.; Innocenti, C.; Sangregorio, C.; Meneghetti, M. Magnetic iron oxide nanoparticles with tunable size and free surface obtained via a "green" approach based on laser irradiation in water. *J. Mater. Chem.* **2011**, *21*, 18665–18673. [CrossRef]

15. Pandey, B.K.; Shahi, A.K.; Shah, J.; Kotnala, R.K.; Gopal, R. Optical and magnetic properties of fe$_2$o$_3$ nanoparticles synthesized by laser ablation/fragmentation technique in different liquid media. *Appl. Surf. Sci.* **2014**, *289*, 462–471. [CrossRef]

16. Svetlichnyi, V.A.; Shabalina, A.V.; Lapin, I.N.; Goncharova, D.A.; Velikanov, D.A.; Sokolov, A.E. Characterization and magnetic properties study for magnetite nanoparticles obtained by pulsed laser ablation in water. *Appl. Phys. A* **2017**, *123*, 763. [CrossRef]

17. Ismail, R.A.; Sulaiman, G.M.; Abdulrahman, S.A.; Marzoog, T.R. Antibacterial activity of magnetic iron oxide nanoparticles synthesized by laser ablation in liquid. *Mater. Sci. Eng. C* **2015**, *53*, 286–297. [CrossRef] [PubMed]

18. Kanitz, A.; Hoppius, J.S.; del Mar Sanz, M.; Maicas, M.; Ostendorf, A.; Gurevich, E.L. Synthesis of magnetic nanoparticles by ultrashort pulsed laser ablation of iron in different liquids. *Chemphyschem* **2017**, *18*, 1155–1164. [CrossRef] [PubMed]

19. Santillán, J.M.J.; Muñetón Arboleda, D.; Coral, D.F.; Fernández van Raap, M.B.; Muraca, D.; Schinca, D.C.; Scaffardi, L.B. Optical and magnetic properties of fe nanoparticles fabricated by femtosecond laser ablation in organic and inorganic solvents. *Chemphyschem* **2017**, *18*, 1192–1209.

20. Jongnam, P.; Eunwoong, L.; Nong-Moon, H.; Misun, K.; Chul, K.S.; Yosun, H.; Je-Geun, P.; Han-Jin, N.; Jae-Young, K.; Jae-Hoon, P.; et al. One-nanometer-scale size-controlled synthesis of monodisperse magnetic iron oxide nanoparticles. *Angew. Chem.* **2005**, *117*, 2932–2937.

21. Franzel, L.; Bertino, M.F.; Huba, Z.J.; Carpenter, E.E. Synthesis of magnetic nanoparticles by pulsed laser ablation. *Appl. Surf. Sci.* **2012**, *261*, 332–336. [CrossRef]

22. Khan, U.; Adeela, N.; Irfan, M.; Ali, H.; Han, X.F. Temperature mediated morphological and magnetic phase transitions of iron/iron oxide core/shell nanostructures. *J. Alloys Compd.* **2017**, *696*, 362–368. [CrossRef]

23. Pardoe, H.; Chua-anusorn, W.; St. Pierre, T.G.; Dobson, J. Structural and magnetic properties of nanoscale iron oxide particles synthesized in the presence of dextran or polyvinyl alcohol. *J. Magn. Magn. Mater.* **2001**, *225*, 41–46. [CrossRef]

24. Qiang, Y.; Antony, J.; Sharma, A.; Nutting, J.; Sikes, D.; Meyer, D. Iron/iron oxide core-shell nanoclusters for biomedical applications. *J. Nanopart. Res.* **2006**, *8*, 489–496. [CrossRef]

25. Sugioka, K.; Cheng, Y. Femtosecond laser three-dimensional micro- and nanofabrication. *Appl. Phys. Rev.* **2014**, *1*, 041303. [CrossRef]

26. Zhang, D.; Chen, F.; Fang, G.; Yang, Q.; Xie, D.; Qiao, G.; Li, W.; Si, J.; Hou, X. Wetting characteristics on hierarchical structures patterned by a femtosecond laser. *J. Micromech. Microeng.* **2010**, *20*, 075029. [CrossRef]

27. Zhang, D.; Chen, F.; Yang, Q.; Si, J.; Hou, X. Mutual wetting transition between isotropic and anisotropic on directional structures fabricated by femtosecond laser. *Soft Matter* **2011**, *7*, 8337–8342. [CrossRef]

28. Zhang, D.; Chen, F.; Yang, Q.; Yong, J.; Bian, H.; Ou, Y.; Si, J.; Meng, X.; Hou, X. A simple way to achieve pattern-dependent tunable adhesion in superhydrophobic surfaces by a femtosecond laser. *ACS Appl. Mater. Interfaces* **2012**, *4*, 4905–4912. [CrossRef] [PubMed]

29. Moussa, S.; Atkinson, G.; El-Shall, M.S. Laser-assisted synthesis of magnetic fe/fe₂o₃ core: Carbon-shell nanoparticles in organic solvents. *J. Nanopart. Res.* **2013**, *15*, 1470. [CrossRef]

30. Zufía-Rivas, J.; Morales, P.; Veintemillas-Verdaguer, S. Effect of the sodium polyacrylate on the magnetite nanoparticles produced by green chemistry routes: Applicability in forward osmosis. *Nanomaterials* **2018**, *8*, 470.

31. Nguyen, V.; Gauthier, M.; Sandre, O. Templated synthesis of magnetic nanoparticles through the self-assembly of polymers and surfactants. *Nanomaterials* **2014**, *4*, 628. [CrossRef] [PubMed]

32. Maneeratanasarn, P.; Khai, T.V.; Kim, S.Y.; Choi, B.G.; Shim, K.B. Synthesis of phase-controlled iron oxide nanoparticles by pulsed laser ablation in different liquid media. *Phys. Status Solidi A* **2013**, *210*, 563–569. [CrossRef]

33. Sukhov, I.A.; Aleksandr, V.S.; Georgii, A.S.; Viau, G.; Garcia, C. Formation of nanoparticles during laser ablation of an iron target in a liquid. *Quantum Electron.* **2012**, *42*, 453. [CrossRef]

34. De Bonis, A.; Lovaglio, T.; Galasso, A.; Santagata, A.; Teghil, R. Iron and iron oxide nanoparticles obtained by ultra-short laser ablation in liquid. *Appl. Surf. Sci.* **2015**, *353*, 433–438. [CrossRef]

35. Marzun, G.; Bönnemann, H.; Lehmann, C.; Spliethoff, B.; Weidenthaler, C.; Barcikowski, S. Role of dissolved and molecular oxygen on cu and ptcu alloy particle structure during laser ablation synthesis in liquids. *Chemphyschem* **2017**, *18*, 1175–1184. [CrossRef] [PubMed]

36. Zhang, D.; Choi, W.; Jakobi, J.; Kalus, M.-R.; Barcikowksi, S.; Cho, S.-H.; Sugioka, K. Spontaneous shape alteration and size separation of surfactant-free silver particles synthesized by laser ablation in acetone during long-period storage. *Nanomaterials* **2018**, *8*, 529. [CrossRef] [PubMed]

37. Zhang, H.; Liang, C.; Liu, J.; Tian, Z.; Shao, G. The formation of onion-like carbon-encapsulated cobalt carbide core/shell nanoparticles by the laser ablation of metallic cobalt in acetone. *Carbon* **2013**, *55*, 108–115. [CrossRef]

38. Robertson, J. Diamond-like amorphous carbon. *Mater. Sci. Eng.* **2002**, *37*, 129–281. [CrossRef]

39. Madrigal-Camacho, M.; Vilchis-Nestor, A.R.; Camacho-López, M.; Camacho-López, M.A. Synthesis of moc@graphite nps by short and ultra-short pulses laser ablation in toluene under n2 atmosphere. *Diamond Relat. Mater.* **2018**, *82*, 63–69. [CrossRef]

40. De Bonis, A.; Santagata, A.; Galasso, A.; Laurita, A.; Teghil, R. Formation of titanium carbide (tic) and tic@c core-shell nanostructures by ultra-short laser ablation of titanium carbide and metallic titanium in liquid. *J. Colloid Interfaces Sci.* **2017**, *489*, 76–84. [CrossRef] [PubMed]

41. Zhang, H.; Liu, J.; Tian, Z.; Ye, Y.; Cai, Y.; Liang, C.; Terabe, K. A general strategy toward transition metal carbide/carbon core/shell nanospheres and their application for supercapacitor electrode. *Carbon* **2016**, *100*, 590–599. [CrossRef]

42. Giorgetti, E.; Muniz-Miranda, M.; Marsili, P.; Scarpellini, D.; Giammanco, F. Stable gold nanoparticles obtained in pure acetone by laser ablation with different wavelengths. *J. Nanopart. Res.* **2012**, *14*, 648. [CrossRef]

43. Safronov, A.; Beketov, I.; Komogortsev, S.; Kurlyandskaya, G.; Medvedev, A.; Leiman, D.; Larrañaga, A.; Bhagat, S. Spherical magnetic nanoparticles fabricated by laser target evaporation. *AIP Adv.* **2013**, *3*, 052135. [CrossRef]

44. Osipov, V.V.; Platonov, V.V.; Uimin, M.A.; Podkin, A.V. Laser synthesis of magnetic iron oxide nanopowders. *Tech. Phys.* **2012**, *57*, 543–549. [CrossRef]

45. Grinstaff, M.W.; Salamon, M.B.; Suslick, K.S. Magnetic properties of amorphous iron. *Phys. Rev. B* **1993**, *48*, 269. [CrossRef]

46. Goya, G.; Berquo, T.; Fonseca, F.; Morales, M. Static and dynamic magnetic properties of spherical magnetite nanoparticles. *J. Appl. Phys.* **2003**, *94*, 3520–3528. [CrossRef]

47. Hadjipanayis, G.; Sellmyer, D.J.; Brandt, B. Rare-earth-rich metallic glasses. I. Magnetic hysteresis. *Phys. Rev. B* **1981**, *23*, 3349. [CrossRef]

48. Teja, A.S.; Koh, P.-Y. Synthesis, properties, and applications of magnetic iron oxide nanoparticles. *Prog. Cryst. Growth Charact. Mater.* **2009**, *55*, 22–45. [CrossRef]

49. Wiedwald, U.; Spasova, M.; Salabas, E.; Ulmeanu, M.; Farle, M.; Frait, Z.; Rodriguez, A.F.; Arvanitis, D.; Sobal, N.; Hilgendorff, M. Ratio of orbital-to-spin magnetic moment in co core-shell nanoparticles. *Phys. Rev. B* **2003**, *68*, 064424. [CrossRef]
50. Han, L.; Wiedwald, U.; Biskupek, J.; Fauth, K.; Kaiser, U.; Ziemann, P. Nanoscaled alloy formation from self-assembled elemental co nanoparticles on top of pt films. *Beilstein J. Nanotechnol.* **2011**, *2*, 473. [CrossRef] [PubMed]
51. Xu, Y.Y.; Rui, X.F.; Fu, Y.Y.; Zhang, H. Magnetic properties of α-fe$_2$o$_3$ nanowires. *Chem. Phys. Lett.* **2005**, *410*, 36–38. [CrossRef]
52. Wu, C.; Yin, P.; Zhu, X.; OuYang, C.; Xie, Y. Synthesis of hematite (α-fe$_2$o$_3$) nanorods: Diameter-size and shape effects on their applications in magnetism, lithium ion battery, and gas sensors. *J. Phys. Chem. B* **2006**, *110*, 17806–17812. [CrossRef] [PubMed]
53. Bødker, F.; Hansen, M.F.; Koch, C.B.; Lefmann, K.; Mørup, S. Magnetic properties of hematite nanoparticles. *Phys. Rev. B* **2000**, *61*, 6826.
54. Bødker, F.; Mørup, S. Size dependence of the properties of hematite nanoparticles. *Europhys. Lett.* **2000**, *52*, 217.
55. Shih, C.-Y.; Streubel, R.; Heberle, J.; Letzel, A.; Shugaev, M.; Wu, C.; Schmidt, M.; Gokce, B.; Barcikowski, S.; Zhigilei, L. Two mechanisms of nanoparticle generation in picosecond laser ablation in liquids: The origin of the bimodal size distribution. *Nanoscale* **2018**, *10*, 6900–6910. [CrossRef] [PubMed]
56. Ibrahimkutty, S.; Wagener, P.; Rolo, T.d.S.; Karpov, D.; Menzel, A.; Baumbach, T.; Barcikowski, S.; Plech, A. A hierarchical view on material formation during pulsed-laser synthesis of nanoparticles in liquid. *Sci. Rep.* **2015**, *5*, 16313. [CrossRef] [PubMed]
57. Semaltianos, N.G.; Friedt, J.-M.; Chassagnon, R.; Moutarlier, V.; Blondeau-Patissier, V.; Combe, G.; Assoul, M.; Monteil, G. Oxide or carbide nanoparticles synthesized by laser ablation of a bulk hf target in liquids and their structural, optical, and dielectric properties. *J. Appl. Phys.* **2016**, *119*, 204903. [CrossRef]
58. Kalus, M.-R.; Barsch, N.; Streubel, R.; Gokce, E.; Barcikowski, S.; Gokce, B. How persistent microbubbles shield nanoparticle productivity in laser synthesis of colloids—Quantification of their volume, dwell dynamics, and gas composition. *Phys. Chem. Chem. Phys.* **2017**, *19*, 7112–7123. [CrossRef] [PubMed]
59. Lauterborn, W.; Vogel, A. Shock wave emission by laser generated bubbles. In *Bubble Dynamics and Shock Waves*; Springer: Berlin, Germany, 2013; pp. 67–103.
60. Suslick, K.S.; Choe, S.-B.; Cichowlas, A.A.; Grinstaff, M.W. Sonochemical synthesis of amorphous iron. *Nature* **1991**, *353*, 414. [CrossRef]
61. Dell'Aglio, M.; Gaudiuso, R.; De Pascale, O.; De Giacomo, A. Mechanisms and processes of pulsed laser ablation in liquids during nanoparticle production. *Appl. Surf. Sci.* **2015**, *348*, 4–9.
62. Lázár, K.; Varga, L.K.; Kovács Kis, V.; Fekete, T.; Klencsár, Z.; Stichleutner, S.; Szabó, L.; Harsányi, I. Electric explosion of steel wires for production of nanoparticles: Reactions with the liquid media. *J. Alloys Compd.* **2018**, *763*, 759–770.
63. Beketov, I.V.; Safronov, A.P.; Bagazeev, A.V.; Larrañaga, A.; Kurlyandskaya, G.V.; Medvedev, A.I. In situ modification of fe and ni magnetic nanopowders produced by the electrical explosion of wire. *J. Alloys Compd.* **2014**, *586*, S483–S488. [CrossRef]
64. Liu, P.; Cui, H.; Yang, G. Synthesis of body-centered cubic carbon nanocrystals. *Cryst. Growth Des.* **2008**, *8*, 581–586. [CrossRef]
65. Seyedeh Zahra, M.; Parviz, P.; Ali, R.; Soghra, M.; Rasoul, S.-B. Generation of various carbon nanostructures in water using ir/uv laser ablation. *J. Phys. D Appl. Phys.* **2013**, *46*, 165303.

nanomaterials

MDPI

Article

Local Melting of Gold Thin Films by Femtosecond Laser-Interference Processing to Generate Nanoparticles on a Source Target

Yoshiki Nakata [1,*], Keiichi Murakawa [1], Noriaki Miyanaga [1], Aiko Narazaki [2], Tatsuya Shoji [3] and Yasuyuki Tsuboi [3]

[1] Institute of Laser Engineering, Osaka University, 2-6 Yamadaoka, Suita, Osaka 565-0871, Japan; murakawa-k@ile.osaka-u.ac.jp (K.M.); miyanaga@ilt.or.jp (N.M.)
[2] National Institute of Advanced Industrial Science and Technology, Central 5, Higashi 1-1-1, Tsukuba, Ibaraki 305-8565, Japan; narazaki-aiko@aist.go.jp
[3] Graduate School of Science, Osaka City University, 3-3-138 Sugimoto Sumiyoshi-ku, Osaka 558-8585, Japan; t-shoji@sci.osaka-cu.ac.jp (T.S.); twoboys@sci.osaka-cu.ac.jp (Y.T.)
*   Correspondence: nakata-y@ile.osaka-u.ac.jp; Tel.: +81-6-6879-8729

Received: 15 June 2018; Accepted: 23 June 2018; Published: 28 June 2018

Abstract: Shape- and size-controlled metallic nanoparticles are very important due to their wide applicability. Such particles have been fabricated by chemosynthesis, chemical-vapor deposition, and laser processing. Pulsed-laser deposition and laser-induced dot transfer use ejections of molten layers and solid-liquid-solid processes to fabricate nanoparticles with a radius of some tens to hundreds of nm. In these processes, the nanoparticles are collected on an acceptor substrate. In the present experiment, we used laser-interference processing of gold thin films, which deposited nanoparticles directly on the source thin film with a yield ratio. A typical nanoparticle had roundness $f_r = 0.99$ and circularity $f_{circ} = 0.869$, and the radius was controllable between 69 and 188 nm. The smallest radius was 82 nm on average, and the smallest standard deviation was 3 nm. The simplicity, high yield, and ideal features of the nanoparticles produced by this method will broaden the range of applications of nanoparticles in fields such as plasmonics.

Keywords: femtosecond laser; interference; laser processing; melt; nanoparticle; gold; thin film

1. Introduction

Shape- and size-controlled metallic nanoparticles have a variety of applications in the fields of plasmonics, catalysts, biology, etc. Such nanoparticles have been fabricated by chemosynthesis [1,2], chemical-vapor deposition [3], electron-beam lithography [4], laser processing, etc. In laser processing, pulsed-laser deposition (PLD) [5–7] and local melting of thin films have been employed, with the latter technique using melting and re-solidification of the film [8–11]. For area processing by a Gaussian beam, the particle sizes are dispersed due to the Rayleigh instability criterion [8]. On the other hand, localization of the melt area by patterned illumination using a mask to avoid the effects of the Rayleigh instability results in a relatively uniform size distribution [9–11]. In this case, the thin film is ablated through the mask, and the rest of the source thin film melts, shrinks, and forms nanoparticles. The laser-interference processing technique has also been applied to metallic thin films. Using this technique, liquid structures such as nanodrops [12–15], nanobumps, and nanowhiskers [13–17] have been fabricated for cases where the thermal-diffusion length is short when compared to the period of the interference pattern. In these processes, motions of the liquid source metal are induced periodically in space by the interference pattern, and nano-sized structures freeze simultaneously after photon and emission of radiation. We call this mechanism the solid-liquid-solid (SLS) [16], in contrast to the vapor-liquid-solid (VLS) mechanism that is used to generate one-dimensional (1D) nanomaterials [18]. In nanowhisker generation, nanodrops detach from the nanowhiskers and remain on the source target, but they blow off before morphological measurements can be carried out. In some other experiments, a single-focus spot has been employed in the laser-induced dot transfer (LIDT) technique [19–22]. This is a variation of the laser-induced forward transfer (LIFT) technique [23–25] where a nanoparticle from a source film is caught on an acceptor substrate.

In the present experiment, we collected gold nanoparticles directly on the source target without requiring an acceptor substrate (Laser-Induced Dot caught On Source target; LIDOS). We measured the morphologies of the nanoparticles using scanning electron microscopy (SEM) without any cleaning of the target. We analyzed the dependence of the size distributions of the nanoparticles on parameters such as the film thickness and fluence. The results that we obtained are supported by a model that explains the formation mechanism.

2. Materials and Methods

Our experimental setup is shown in Figure 1. We used a femtosecond laser beam (IFRIT, Cyber Laser, Tokyo, Japan) operating at 785 nm with a 240 fs (full-width at half-maximum, FWHM) pulse width. The beam was split by a diffractive optical element, which was optimized to the wavelength (the diffraction efficiency was approximately 60%) to generate four first-order diffracted beams. The four beams were focused on the surface of the target via a demagnification system, for which the focal lengths of the achromatic lenses were $f_1 = 300$ mm and $f_2 = 40$ mm. The original beam diameter was 6.0 mm (FWHM at $1/e^2$), which was de-magnified to 0.8 mm. The zeroth-order beam was dumped. The correlation angle was $\theta = 16.7°$, and the period of the interference pattern was $\Lambda = 19.3$ μm. For this experiment, we chose a gold thin-film target, which has surface-plasmon resonance [26] and stability. We deposited the films on silica-glass substrates by magnetron sputtering and was electrically insulated from the earth. All of the experiments were performed using a single shot of laser irradiation at atmospheric conditions. We imaged the surface structures of the films using SEM (JEOL, JSM-7400FS, Tokyo, Japan).

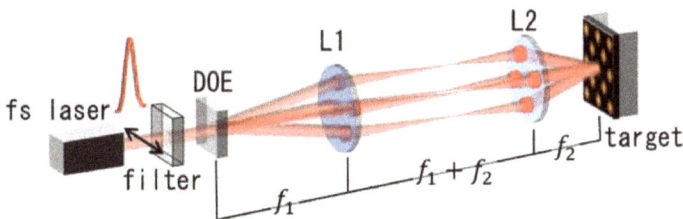

Figure 1. Experimental setup. The pulse width of the laser was 240 fs and the wavelength was 785 nm. DOE (diffractive-optical element) split four 1st order diffracted beams. $f_1 = 300$ mm and $f_2 = 40$ mm.

3. Results and Discussion

Figure 2 shows a top view of the structures fabricated with different film thickness. The target-film thickness, the average fluence over the interference pattern, the average radius and its standard deviation (s.d.) are summarized in Table 1. The precise lattice structures with $\Lambda = 19.3$ μm period were determined by the interference pattern. For the thinner film of 40 nm thickness, the nanoparticles were found to lie on the source target at random locations, as shown in Figure 2a-1. The circular black holes are the areas from which the particles were ejected and through which the substrate surface appeared. Figure 2a-2 is a magnified view of Figure 2a-2, and demonstrates that particles with good roundness could be fabricated. The average radius of the nanoparticles was 188 nm, and the standard deviation was 7 nm. The yield ratio, which is the number of nanoparticles relative to the number of holes, was 83% in the field of view in Figure 2a-1. Figure 3 shows a high-resolution SEM (HR-SEM) image of the same field with 100,000× magnification. The surface of the nanoparticle was smooth, and some nanostructures smaller than 10 nm adhered to the surface; these may have been smaller nanoparticles that had condensed from the gold vapor. The radius of this nanoparticle was 174.8 nm. The circularity f_{circ} and roundness f_r of the nanoparticle were defined by the following equations:

$$f_{circ} = 4\pi S / P^2 \tag{1}$$

$$f_r = 4S / (2a)^2, \tag{2}$$

where S is the surface area of the nanoparticle; P is the perimeter; and $2a$ is the length of the major axis, assuming the shape to be an ellipsoid. For this nanoparticle, we obtained $f_{circ} = 0.869$ and $f_r = 0.99$, so it was a fairly round sphere with a slightly rough surface. Note that f_{circ} is affected by the resolution of the SEM.

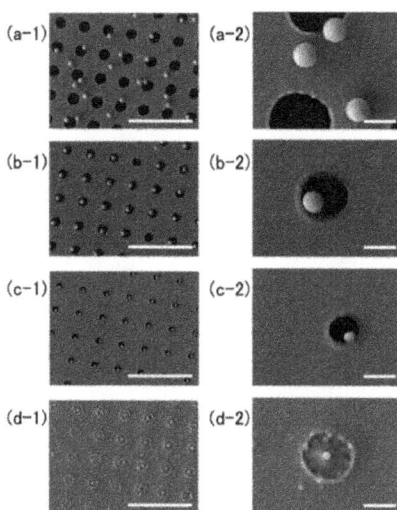

Figure 2. Scanning electron microscopy (SEM) images of the target surface with different film thickness. (a-1) 40 nm; (b-1) 50 nm; (c-1) 60 nm; and (d-1) 100 nm, and their magnified images on right column: (a-2) 40 nm; (b-2) 50 nm; (c-2) 60 nm; and (d-2) 100 nm. The experimental parameters and results are summarized in Table 1. The white bars in the images on the left and right columns represent 5 μm and 500 nm, respectively.

Table 1. Parameters and results for the experiments shown in Figure 2, Figure 3, and Figure 6. In the case of Figure 2c, nanoparticles should adhere to the edge of holes of the film and not be separated. In the case of Figure 2d, the nano-structure was nano-projecting, not nanoparticle.

Parameter	Figure 2(a-1,a-2), Figure 3	Figure 2(b-1,b-2)	Figure 2(c-1,c-2)	Figure 2(d-1,d-2)	Figure 6(a-1,a-2)	Figure 6(b-1,b-2)	Figure 6(c-1,c-2)
film thickness (nm)	40	50	60	100	30	40	40
fluence (mJ/cm²)	73.2	73.2	73.2	73.2	58.6	58.6	169.9
averaged radius (nm)	188	170	82	69	n.d.	137	138
standard d. (nm)	7	3	5	9	n.d.	10	27

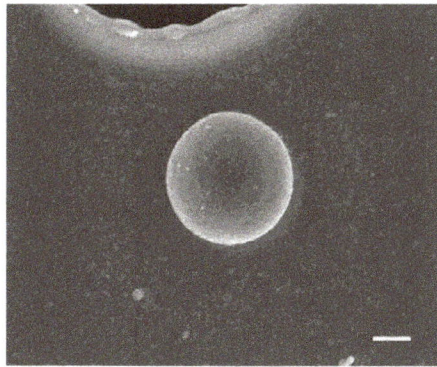

Figure 3. Observation with high-resolution scanning electron microscopy (HR-SEM) of a single nanoparticle shown in Figure 2a-1. The film thickness was 40 nm, and the fluence was 73.2 mJ/cm^2. The radius was 175 nm. The white bar in the image represent 100 nm.

When the film thickness was 50 nm, smaller nanoparticles were fabricated, as shown in Figure 2(b-1,b-2). Most remained on the lower left sides of the holes, but the reason for this is not known. The yield ratio was 93%. The standard deviation of the radius was 3 nm, which was the smallest dispersion in this experiment. When the film thickness was 60 nm, smaller nanoparticles always remained on the lower right edges of the holes, as shown in Figure 2(c-1,c-2). The yield ratio of the field of view was 100%. When the thickest film of 100 nm was used, only the textured surface of the gold film was seen, as shown in Figure 2(d-1,d-2). White spots, which show the existence of projecting structures, could be seen at the centers of the depressions in the processed area. Owing to the thickness of this film, none of these depressions punched through to expose the substrate.

The radius as a function of film thickness is summarized in Figure 4. The smallest film thickness was 40 nm. The radius of the nanoparticles was smaller than 200 nm in all cases. It is apparent that the radius of the nanoparticles decreased as the film thickness increased from 40 to 60 nm. The results that we obtained are supported by a model that explains the formation mechanism, as shown in Figure 5 [16,27,28]. For every thickness, the films were almost opaque at 785 nm. As the thermal conductivity of gold at room temperature is far higher than that of silica glass, the temperature distribution along the substrate is normal inside the gold film levels immediately. As a result, the rise in temperature is inversely proportional to the mass of the film, as illustrated in the top curves in the figure. For the thinner film shown in Figure 5a, the region at temperatures above the melting point ($T_{m.p.} = 1064\,^\circ\text{C}$) was wider than that for the thicker films, as illustrated in Figure 5b,c. The molten layer was launched from the substrate by the reaction to volumetric expansion during the solid-to-liquid phase transition, as shown in Figure 5a-i. It was then squeezed and formed a nanoparticle due to surface tension, as in Figure 5a-ii. It solidified because of cooling by photon emission, and could then be pulled by electrostatic forces and adhered to the target surface, as shown in Figure 5a-iii. Most of the nanoparticles adhered to the metal surface because of the metallic bond. However, when the film was too thin, as shown in Figure 6(a-1,a-2), the region simply boiled away.

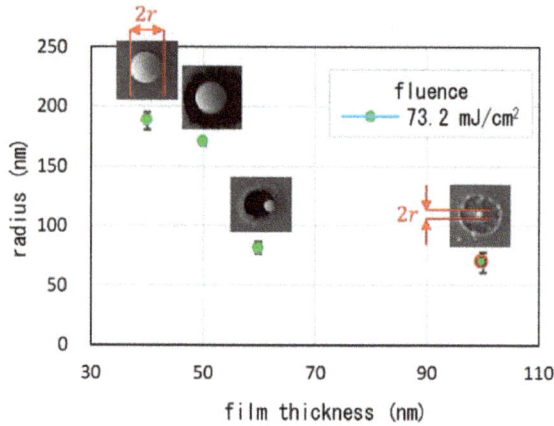

Figure 4. The radii of the nanoparticles as a function of film thickness. The fluence was 73.2 mJ/cm^2. The error bars reflect the s.d.

For the film thickness of 50 nm, the temperature was lower, which resulted in a less-effective launch, as shown in Figure 5b-i. A nanoparticle is formed at the top center of the molten layer because of surface tension, as illustrated in Figure 5b-ii. If it freezes at this time, it can form a nanodrop on a hollow bump [12], or a nanowhisker [16,17], which is left after the nanoparticle detaches. The resulting nanoparticles are left on the processed area or squeezed to the edges of the holes by surface tension, as illustrated in Figure 5b-iii and shown in Figure 2b,c. For a film thickness of 100 nm, a hole could not be punched through the gold film at the given fluence, as shown in Figure 2(d-1,d-2). In this case, the motion of the molten layer resulted in a textured film surface, as shown in Figure 5c-i–c-iii.

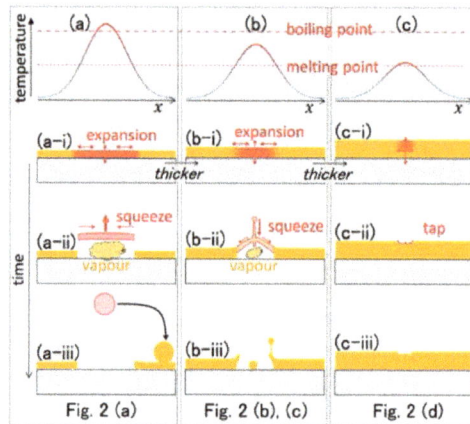

Figure 5. Schematic illustration of the solid-liquid-solid (SLS) process. Each panel illustrates the temporal change of a spot in an interference pattern. The correspondence with the results shown in Figure 2 is noted on the bottom. From (**a–c**), the thickness of the film is listed as following: (**a**), 40 nm; (**b**), 50 and 60 nm; (**c**), 100 nm.

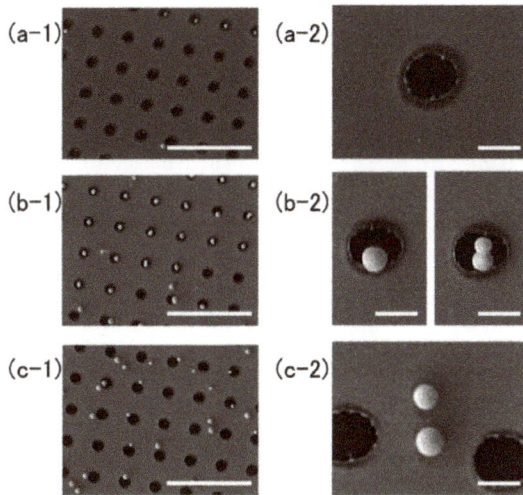

Figure 6. SEM images of the target surface. (**a-1**) film thickness was 30 nm, and fluence was 58.6 mJ/cm^2; (**b-1**) film thickness was 40 nm, and fluence was 58.6 mJ/cm^2; (**c-1**) film thickness was 40 nm, and fluence was 169.9 mJ/cm^2. Pictures on the right column are the corresponding magnified images: (**a-2**) film thickness was 30 nm, and fluence was 58.6 mJ/cm^2; (**b-2**) film thickness was 40 nm, and fluence was 58.6 mJ/cm^2; (**c-2**) film thickness was 40 nm, and fluence was 169.9 mJ/cm^2. The experimental parameters and results are summarized in Table 1. The white bars in the images on the left and right columns represent 5 μm and 500 nm, respectively.

The radius of the nanoparticles as a function of the fluence are summarized in Figure 7 for a film thickness of 40 nm. The radius was smaller than 200 nm in all cases. It is interesting that mixed nanoparticles such as singles and twins were produced, as shown in Figure 6(b-1,b-2). In Figure 7, data for single particles are plotted. It is interesting that the formation of twins or multiple droplets in a string is seen also in the behavior of water solutions [29]. The process of drop detachment was controlled in those experiments by dissolving polymeric molecules, thus changing the viscoelasticity of the solution. In our case, it may be possible to control multiple-nanoparticle formation by changing the material, e.g., using silver, chromium and alloys to obtain different values of viscoelasticity. When the fluence was at the highest value that we employed, 169.9 mJ/cm^2, the process caused more spattering. As can be seen in Figure 6(c-1,c-2), the standard deviation of the radius was highest, 27 nm.

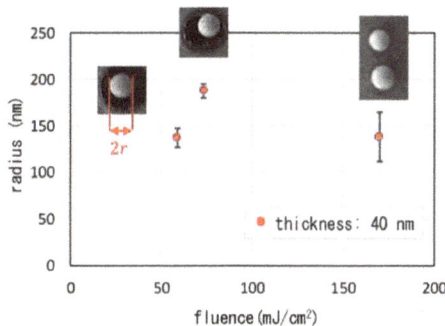

Figure 7. Radius of nanoparticles as a function of fluence. The film thickness was 40 nm. The error bars reflect the s.d.

Nanomaterials **2018**, *8*, 477

We now compare this method with others. The roundness of the nanoparticles produced in this experiment was excellent, and the radius controllable between 69 and 188 nm. In PLD, the radii of the nanoparticles that condensed from the gas phase ranged from a few nm to some tens of nm [5,6]. On the other hand, the droplets fabricated by PLD were some hundreds of nm in radius [7]. Using LIDT, nanoparticles with radii smaller than 250 nm can be fabricated [19,30]. Electron-beam lithography is utilized for fabricating aligned nanoparticles a few nm in diameter [4]. Chemosynthesis can be used to fabricate nanoparticles and nanorods with radii of a few nm in both structures [1]. In summary, this technique is a good alternative to PLD or LIDT, and can generate nanoparticles with radii of tens or hundreds of nm via the SLS process.

4. Conclusions

Using the SLS process, we successfully fabricated gold nanoparticles with radii of tens to hundreds of nm and good roundness-which were subsequently deposited on the source substrate-using irradiation through the interference pattern of a fs laser. The film thickness and fluence were the key parameters for controlling the sizes of the nanoparticles, with a thinner film resulting in a larger nanoparticle radius. The smallest radius was 82 nm on average, and the smallest standard deviation was 3 nm. A typical nanoparticle was found to have roundness $f_r = 0.99$.

Compared with the methods such as chemosynthesis, VLS and PLD, SLS is useful for fabricating pure and uniform nanomaterials. No catalyst or chemosynthetic solution is required, and a more uniform size distribution will result in better plasmonic resonance properties. We anticipate that different source materials such as metals, alloys, and non-metals with plasticity can produce nanoparticles using the SLS mechanism. The process is very simple and does not require cleaning, temperature control, evacuation, purification, etc. These advantages will broaden the range of applications for such nanoparticles.

Author Contributions: Conceptualization, Y.N.; Methodology, Y.N.; Validation, Y.N.; Formal Analysis, Y.N. Investigation, K.M.; Resources, N.M.; Data Curation, K.M. and Y.N.; Writing-Original Draft Preparation, Y.N.; Writing-Review & Editing, Y.N.; Visualization, Y.N.; Supervision, Y.N.; Project Administration, Y.N.; Funding Acquisition, Y.N., A.N., T.S., Y.T.

Funding: This research was funded by the Japan Society for the Promotion of Science (JSPS) (23360035, A16H038850).

Conflicts of Interest: The authors declare no conflict of interest.

References

1. Jana, N.R.; Gearheart, L.; Murphy, C.J. Wet Chemical Synthesis of High Aspect Ratio Cylindrical Gold Nanorods. *J. Phys. Chem. B* **2001**, *105*, 4065–4067. [CrossRef]
2. Silver, P.; Pietrobon, B.; Mceachran, M.; Kitaev, V. Synthesis of Size-Controlled Faceted Tunable Plasmonic Properties and Self- Assembly of These Nanorods. *ACS Nano* **2009**, *3*, 21–26. [CrossRef]
3. Okumura, M.; Tsubota, S.; Iwamoto, M.; Haruta, M. Chemical Vapor Deposition of Gold Nanoparticles on MCM-41 and Their Catalytic Activities for the Low-temperature Oxidation of CO and of H_2. *Chem. Lett.* **1998**, *27*, 315–316. [CrossRef]
4. Vieu, C.; Carcenac, F.; Pepin, A.; Chen, Y.; Mejias, M.; Lebib, A.; Manin-Ferlazzo, L.; Couraud, L.; Launois, H. Electron beam lithography: Resolution limits and applications. *Appl. Surf. Sci.* **2000**, *164*, 111–117. [CrossRef]
5. Guczi, L.; Horváth, D.; Pászti, Z.; Tóth, L.; Horváth, Z.E.; Karacs, A.; Petõ, G. Modeling Gold Nanoparticles: Morphology, Electron Structure, and Catalytic Activity in CO Oxidation. *J. Phys. Chem. B* **2000**, *104*, 3183–3193. [CrossRef]
6. Muramoto, J.; Sakamoto, I.; Nakata, Y.; Okada, T.; Maeda, M. Influence of electric field on the behavior of Si nanoparticles generated by laser ablation. *Appl. Phys. Lett.* **1999**, *75*, 751–753. [CrossRef]
7. Uetsuhara, H.; Goto, S.; Nakata, Y.; Vasa, N.; Okada, T.; Maeda, M. Fabrication of a Ti:sapphire planar waveguide by Pulsed Laser Deposition. *Appl. Phys. A Mater. Sci. Process.* **1999**, *69*, S719–S722. [CrossRef]

8.   Henley, S.J.; Carey, J.D.; Silva, S.R.P. Pulsed-laser-induced nanoscale island formation in thin metal-on-oxide films. *Phys. Rev. B Condens. Matter Mater. Phys.* **2005**, *72*, 195408. [CrossRef]

9.   Mäder, M.; Höche, T.; Gerlach, J.W.; Böhme, R.; Zimmer, K.; Rauschenbach, B. Large area metal dot matrices made by diffraction mask projection laser ablation. *Phys. Status Solidi Rapid Res. Lett.* **2008**, *2*, 34–36. [CrossRef]

10.   Höche, T.; Böhme, R.; Gerlach, J.W.; Rauschenbach, B.; Syrowatka, F. Nanoscale laser patterning of thin gold films. *Philos. Mag. Lett.* **2006**, *86*, 661–667. [CrossRef]

11.   Nakata, Y.; Okada, T.; Maeda, M. Holographic fabrication of micron structures using interfered femtosecond laser beams split by diffractive optics. *Proc. SPIE Int. Soc. Opt. Eng.* **2003**, *4977*. [CrossRef]

12.   Nakata, Y.; Okada, T.; Maeda, M. Nano-sized hollow bump array generated by single femtosecond laser pulse. *Jpn. J. Appl. Phys.* **2003**, *42*, L1452–L1454. [CrossRef]

13.   Nakata, Y.; Momoo, K.; Hiromoto, T.; Miyanaga, N. Generation of superfine structure smaller than 10 nm by interfering femtosecond laser processing. In Proceedings of the SPIE the International Society for Optical Engineering, San Francisco, CA, USA, 22–27 January 2011; Volume 7920.

14.   Nakata, Y.; Tsuchida, K.; Miyanaga, N.; Furusho, H. Liquidly process in femtosecond laser processing. *Appl. Surf. Sci.* **2009**, *255*, 9761–9763. [CrossRef]

15.   Nakata, Y. Frozen water drops in the nanoworld. *SPIE Newsroom* **2009**, *2*, 1–2. [CrossRef]

16.   Nakata, Y.; Miyanaga, N.; Momoo, K.; Hiromoto, T. Solid-liquid-solid process for forming free-standing gold nanowhisker superlattice by interfering femtosecond laser irradiation. *Appl. Surf. Sci.* **2013**, *274*, 27–32. [CrossRef]

17.   Nakata, Y.; Miyanaga, N.; Momoo, K.; Hiromoto, T. Template free synthesis of free-standing silver nanowhisker and nanocrown superlattice by interfering femtosecond laser irradiation Template free synthesis of free-standing silver nanowhisker and nanocrown superlattice by interfering femtosecond laser irradiation. *Jpn. J. Appl. Phys.* **2014**, *53*, 096701.

18.   Wagner, R.S.; Ellis, W.C. Vapor-liquid-solid mechanism of single crystal growth. *Appl. Phys. Lett.* **1964**, *4*, 89–90. [CrossRef]

19.   Narazaki, A.; Sato, T.; Kurosaki, R.; Kawaguchi, Y.; Niino, H. Nano- and microdot array formation of FeSi2 by nanosecond excimer laser-induced forward transfer. *Appl. Phys. Express* **2008**, *1*, 0570011–0570013. [CrossRef]

20.   Narazaki, A.; Sato, T.; Kurosaki, R.; Kawaguchi, Y.; Niino, H. Nano- and microdot array formation by laser-induced dot transfer. *Appl. Surf. Sci.* **2009**, *255*, 9703–9706. [CrossRef]

21.   Kuznetsov, A.I.; Evlyukhin, A.B.; Reinhardt, C.; Seidel, A.; Kiyan, R.; Cheng, W.; Ovsianikov, A.; Chichkov, B.N. Laser-induced transfer of metallic nanodroplets for plasmonics and metamaterial applications. *J. Opt. Soc. Am. B* **2009**, *26*, B130. [CrossRef]

22.   Willis, D.A.; Grosu, V. Microdroplet deposition by laser-induced forward transfer. *Appl. Phys. Lett.* **2005**, *86*, 1–3. [CrossRef]

23.   Levene, M.L.; Scott, R.D.; Siryj, B.W. Material Transfer Recording. *Appl. Opt.* **1970**, *9*, 2260. [CrossRef] [PubMed]

24.   Bohandy, J. Metal deposition from a supported metal film using an excimer laser. *J. Appl. Phys.* **1986**, *60*, 10–12. [CrossRef]

25.   Nakata, Y.; Okada, T. Time-resolved microscopic imaging of the laser-induced forward transfer process. *Appl. Phys. A Mater. Sci. Process.* **1999**, *69*, S275–S278. [CrossRef]

26.   Jain, P.; Lee, K.; El-Sayed, I.; El-Sayed, M. Calculated absorption and scattering properties of gold nanoparticles of different size, shape, and composition: Applications in biological imaging and biomedicine. *J. Phys. Chem. B* **2006**, *110*, 7238–7248. [CrossRef] [PubMed]

27.   Wu, C.; Zhigilei, L.V. Nanocrystalline and Polyicosahedral Structure of a Nanospike Generated on Metal Surface Irradiated by a Single Femtosecond Laser Pulse. *J. Phys. Chem. C* **2016**, *120*, 4438–4447. [CrossRef]

28.   Ivanov, D.S.; Lin, Z.; Rethfeld, B.; O'Connor, G.M.; Glynn, T.J.; Zhigilei, L.V. Nanocrystalline structure of nanobump generated by localized photoexcitation of metal film. *J. Appl. Phys.* **2010**, *107*. [CrossRef]

29. Clasen, C.; Bico, J.; Entov, V.M.; McKinley, G.H. "Gobbling drops": The jettingdripping transition in flows of polymer solutions. *J. Fluid Mech.* **2009**, *636*, 5–40. [CrossRef]

30. Zhigunov, D.M.; Evlyukhin, A.B.; Shalin, A.S.; Zywietz, U.; Chichkov, B.N. Femtosecond Laser Printing of Single Ge and SiGe Nanoparticles with Electric and Magnetic Optical Resonances. *ACS Photonics* **2018**, *5*, 977–983. [CrossRef]

nanomaterials

MDPI

Article

Dual THz Wave and X-ray Generation from a Water Film under Femtosecond Laser Excitation

Hsin-hui Huang [1], Takeshi Nagashima [2,*], Wei-hung Hsu [1], Saulius Juodkazis [3,4,*] and Koji Hatanaka [1,5,6,*]

[1] Research Center for Applied Sciences, Academia Sinica, Taipei 115, Taiwan;
hsinhuih@gate.sinica.edu.tw (H.-h.H.); jacky81418@gate.sinica.edu.tw (W.-h.-H.)

[2] Faculty of Science and Engineering, Setsunan University, 17-8 Ikeda-Nakamachi, Neyagawa, Osaka 572-8508, Japan

[3] Nanotechnology Facility, Center for Micro-Photonics, Swinburne University of Technology, Hawthorn, VIC 3122, Australia

[4] Melbourne Centre for Nanofabrication, the Victorian Node of the Australian National Fabrication Facility, Clayton, VIC 3168, Australia

[5] College of Engineering, Chang Gung University, Taoyuan 33302, Taiwan

[6] Department of Materials Science and Engineering, National Dong-Hwa University, Hualien 97401, Taiwan

[*] Correspondence: t-nagash@mpg.setsunan.ac.jp (T.N.); sjuodkazis@swin.edu.au (S.J.);
kojihtnk@gate.sinica.edu.tw (K.H.); Tel.: +81-72-813-1167 (T.N.); +61-3-9214-8718 (S.J.);
+886-2-2787-3132 (K.H.)

Received: 25 June 2018; Accepted: 11 July 2018; Published: 13 July 2018

Abstract: Simultaneous emission of the THz wave and hard X-ray from thin water free-flow was induced by the irradiation of tightly-focused femtosecond laser pulses (35 fs, 800 nm, 500 Hz) in air. Intensity measurements of the THz wave and X-ray were carried out at the same time with time-domain spectroscopy (TDS) based on electro-optic sampling with a ZnTe(110) crystal and a Geiger counter, respectively. Intensity profiles of the THz wave and X-ray emission as a function of the solution flow position along the incident laser axis at the laser focus show that the profile width of the THz wave is broader than that of the X-ray. Furthermore, the profiles of the THz wave measured in reflection and transmission directions show different features and indicate that THz wave emission is, under single-pulse excitation, induced mainly in laser-induced plasma on the water flow surface. Under double-pulse excitation with a time separation of 4.6 ns, 5–10 times enhancements of THz wave emission were observed. Such dual light sources can be used to characterise materials, as well as to reveal the sequence of material modifications under intense laser pulses.

Keywords: femtosecond laser; intense laser; water; THz wave; time-domain spectroscopy; X-ray; ablation; double-pulse excitation; plasma; z-scan; intensity enhancement

1. Introduction

An intense ($>10^{13}$ W/cm^2) femtosecond laser at near-IR (photon energy $\sim$1 eV) and matter interaction [1–3] induce highly-nonlinear optical processes and result in photon conversion to an X-ray of several-keV [4], as well as to a THz wave at meV [5] in association with the white light continuum in the visible spectral range [6]. Strong laser ablation of solid and solution targets also occurs, which is usually associated with the production of laser plasma [7]. Studies of such photon conversion mechanisms for different applications [8–10] have been carried out separately for widely different wavelengths. Experimental techniques based on the X-ray or THz wave have been contributing immensely to basic physics and chemistry/material science [10–13], have expanded the understanding of the basic mechanisms of intense laser-matter interaction and have contributed to laser-material processing/printing [14].

Gases [15–18], atomic clusters [19–21] and solids [22,23] have been used as targets for THz wave emission from the laser-induced plasmas. Fast electron motions accelerated by laser ponderomotive forces [15,18,19] and the four-mixing process induced by two-colour excitation [17] have been proposed as THz emission mechanisms from laser-induced plasmas. Although such laser-induced plasmas of the solid targets generate a powerful THz wave because of the high plasma densities, there are problems with the instability of the intensity and a limitation in the successive delivery of laser irradiation to the target. The use of liquids or solutions as targets is promising since solutions have moderate atomic density ($\sim 10^{22}$ cm^{-3}) and ease of the delivery, but there have been no reports with liquid targets (water) until recently [5,24]. As for intense laser-induced X-ray emission, it has been widely studied with solids [15,25–28] and solutions [29–38]. One advantage of using solutions, e.g., water, as a target sample is that a fresh and smooth surface can be easily prepared for each laser irradiation even under intense laser irradiation at high repetition rates. Furthermore, the addition of solute chemical agents such as electrolyte, CsCl for instance [39], with different concentrations or nano-particles, gold for instance [40], with different shapes and sizes into water makes it possible to explore different excitation/absorption mechanisms and to control the spectral characteristics of X-ray sources by utilizing solute-dependent monochromatic characteristic X-ray lines and broad components mainly based on bremsstrahlung.

Simultaneous emission of the THz wave and X-ray and combined usages of THz wave and X-ray pulses have recently been introduced as readily-available table-top realisations [41]. To date, experiments on such simultaneous emission with a Al-coated glass substrate [15] and gas targets, He [15] or Ar [20], in vacuum chambers were reported. This type of work, though the numbers of such reports are still quite limited, will be the first step for the discussion of the conversion of near-IR (eV) laser photon energy into photons at the two ends of the spectrum, several-keV (X-ray) and meV (THz wave), and will contribute well to other fields like material sciences. One important parameter for the synchronized emission of the X-ray and THz wave for practical applications is their intensities. It has been reported that chirped femtosecond laser pulses enhance THz wave emission from water [24] and the X-ray from aqueous solutions [42]. In cases of X-ray emission from aqueous solutions, double-pulse excitations [37,43] and the addition of solutes such as electrolytes [37] or gold nano-particles [40] to water are also reported to be effective for the X-ray intensity enhancements. For THz wave emission, one recent experimental study with double-pulse excitation to gas-clusters in a vacuum chamber [21] was reported, but the inter-pulse delay time was limited to 0.5 ns. One advantage of the solution targets, in addition to that described above, is the possibility of a transient solution surface roughening from its original nano-smooth surface by a pre-pulse irradiation. Dynamic changes induced on the liquid surface by plasma formation, capillary transient surface roughening instabilities and mist/droplet formation associated with shock-wave expansion [44] contribute favourably to increased interaction volume and augmented X-ray intensity up to more than an order of magnitude [37,43].

In this study, a simultaneous emission of the THz wave and hard X-ray in air using distilled water as a target irradiated by tightly-focused near-IR femtosecond laser pulses and THz wave emission enhancements under double-pulse excitation are presented.

2. Experimental Section

The experimental setup is shown in Figure 1a. A flat solution flow of distilled water with a thickness smaller than 20 μm was prepared using a metal nozzle (Flatjet Nozzle LARGE, Metaheuristic, Okayama, Japan), and the flow rate was regulated by a circulation pump (PMD-211, SANSO, Hyogo, Japan) controlled by a conventional voltage regulator. The flow rate was set at <70 mL/min. The nozzle was mounted to a rotational and 3D-automatic stages (KS701-20LMS, Suruga Seiki, Shizuoka, Japan), and its position was finely set by a home-made LabView code, as reported previously [43]. Transform-limited femtosecond laser pulses (λ = 800 nm, >35 fs, 1 kHz, linearly-polarized, Mantis, Legend Elite HE USP, Coherent, Inc., Santa Clara, CA, USA) were separated as two beams with different polarizations by a half-wave plate (65-906, Edmund Optics, Barrington, NJ, USA) and a polarization beam splitter (47-048, Edmund Optics, Barrington, NJ, USA). Horizontally- and vertically-polarized pulses were defined as the excitation pulse for X-ray/THz wave generation and the probe pulse for the time-domain spectroscopy (TDS) for THz wave measurements [45,46], respectively. The repetition rate of the excitation pulses was modulated by a wheel chopper (500 Hz, 3502 Optical Chopper, New Focus, CA, USA) for TDS measurements, and the pulses were tightly-focused in air onto the solution flow surface by using an off-axis parabolic mirror (OAPM, 1-inch diameter, the effective focus length $f = 50.8$ mm, the reflection angle of 90 degrees and numerical aperture $NA = 0.25$, 47-097, Edmund Optics, Santa Clara, CA, USA). The incident angle of the excitation pulses along the Z -axis to the solution normal was fixed at 60 degrees for the highest X-ray emission, as reported previously [42]. Under these experimental conditions, each excitation pulse at a 500-Hz repetition rate irradiates the fresh and flat solution flow. Experiments on double-pulse excitations with a pre-pulse (vertically-polarized, 0.1 mJ/pulse, 4.6 ns in advance of the main excitation pulse) were also carried out using an optical delay line (SGSP46-800, Sigma Koki, Tokyo, Japan).

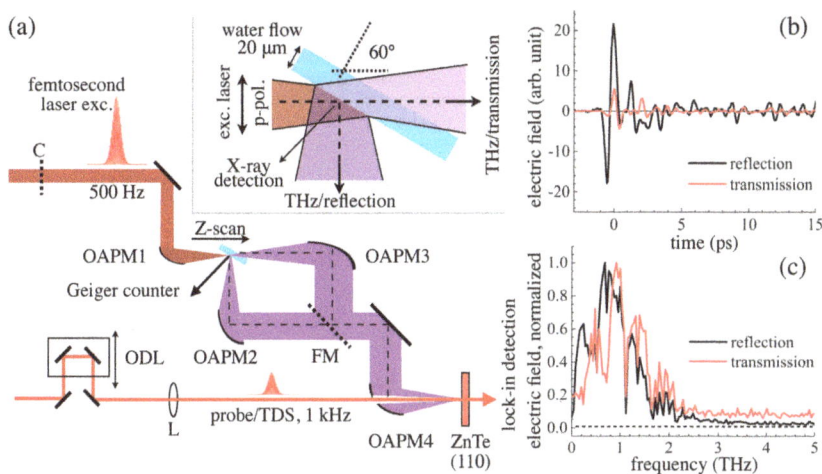

Figure 1. (a) The experimental setup for the simultaneous measurements of THz wave (time-domain spectroscopy (TDS)) and X-ray (Geiger counter). The femtosecond laser pulses (<35 fs, 800 nm, 500 Hz, horizontally-polarised to the solution flow surface, i.e., p-pol.) were focused onto the solution flow; ODL, optical delay line for TDS, L-plano-convex lens ($f = 50$ cm). The thickness of ZnTe(110) crystal for TDS was 1 mm; C, optical chopper (500 Hz) for TDS measurements. The parent focal lengths for off-axis parabolic mirrors (OAPMs) (off-axis parabolic mirrors) were $f = 50.8$ mm (OAPM1, 1-inch diameter), 101.6 mm (OAPM2, 2-inch), 152.4 mm (OAPM3, 2-inch) and 101.6 mm (OAPM4, 2-inch with a hole in its centre for the probe), respectively. The distance between the laser focus and Geiger counter was 12 cm. FM, flip-folding mirror for TDS measurements in the transmission direction. The inset shows the close-up of the solution surface, the laser-water interaction region; (b) Representative TDS signals from water flow when the laser intensity is 0.4 mJ/pulse in reflection and transmission directions at the Z-position for the highest X-ray intensity and (c) their normalized spectra.

X-ray intensity was measured by a Geiger counter (SS315, Radhound, southern scientific, West Sussex, UK). All the measurements were carried out in air under atmospheric pressure (1 atm) at room temperature (RT = 296 K). Its observation angle was 15 degrees to the solution normal towards the excitation side, and its distance from the laser focus was 12 cm. Therefore, it is certain that the Geiger counter detects only X-ray, neither α- nor β-ray. THz wave signals were collected in reflection (30 degrees to the solution normal, 90 degrees to the laser incident direction) and transmission (along the excitation Z-axis) directions with two independent OAPMs (the reflected focal length

$f = 101.6$ mm and 152.4 mm, the off-axis angle of 90 degrees, MPD249-M01, MPD269-M01, ThorLabs, Newton, NJ, USA). As conventional TDS measurements, THz wave and the probe pulses after n variable optical delay (TSDM60-20, OptoSigma, Tokyo, Japan) were focused to a 1 mm-thick ZnTe(110) crystal (Nippon mining & metals Co., Ltd., Tokyo, Japan) by an OAPM (the reflected focal length $f = 101.6$ mm, the off-axis angle of 90 degrees, MPD249H-M01, ThorLabs, Newton, NJ, USA) and a plano-convex lens ($f = 50$ cm), respectively. Lock-in measurements were carried out with a balanced photo-diode (Model2307, New Focus, CA, USA) and a lock-in amplifier (SR830, Stanford Research System, Sunnyvale, CA, USA). For measurements in the transmission direction, an additional flip-folding mirror (FM) was set for the THz wave path to be bent; therefore, the total number of metal mirrors for the transmission measurements was larger by one compared to that for the reflection.

One representative THz wave signal of water in reflection/transmission directions by the EO sampling in TDS measurements is shown in Figure 1b. Detection efficiencies for the reflection and the transmission directions, though their optical paths were shared partly, have not been calibrated; therefore, the THz wave intensities and converted FFT spectra are not comparable quantitatively between the signals in the two observation directions. A single cycle of the electric field oscillation was clearly observed, and the vibration structure afterwards was also very distinct. This is reflected in their Fourier-transformed emission spectra shown in Figure 1c as absorption bands at 1.1 and 1.4 THz due to water vapour, as reported elsewhere [47]. Note that the spectra shown in Figure 1c are normalized. This absorption is considered to be due to the water vapour in the atmosphere of the laboratory and long-living micro-droplets (mist) formation in the vicinity of the water film induced by the laser irradiation every 2 ms. The central frequencies of the observed THz wave were around 0.9–1 THz for the reflection and the transmission. As discussed in detail below, the THz wave was emitted from the area in the vicinity of the upstream-side of the air/water interface. In the transmission, the THz wave transmitted through the water film. Water absorption in the THz wave region was well studied recently, and a transmission experiment of the THz wave through 0.5 mm-thick water films was recently demonstrated using TDS for precise measurements of temperature [48], the discussion of which can be used for characterization of light-water interaction at high intensities. Under conventional transmission conditions thorough 20 μm-thick water toward the transmission direction, but with the incident angle at 60 degrees, the transmittance of the THz wave electric field intensity at 1 THz can be estimated to be 0.64. Since water shows higher absorption at higher frequencies [48,49], the high frequency components in the transmission are expected to be reduced compared with those in the reflection. However, the observed spectra in the transmission showed a slight blue shift compared to that in the reflection. It should be pointed out that the small shift can be due to an extrinsic effect since the spectra obtained in the transmission and the reflection directions have not been calibrated in this study. In the following, the relative intensities of the THz wave obtained in the identical direction are discussed.

Under these conditions, X-ray and THz wave emission was induced at the same time and measured simultaneously. Experiments were performed with different excitation laser intensities, different solution positions along the Z-axis and under the double-pulse excitation. All the experiments were carried out at atmospheric pressure at RT conditions.

3. Results and Discussion

3.1. Laser Intensity Dependences

Figure 2 shows the intensities of the X-ray and THz wave in the reflection and the transmission as a function of the excitation laser intensity. The solution flow position along the Z-axis was optimized for the highest X-ray intensity for each laser intensity. The highest intensity of X-ray was obtained when the highest electron temperature, T_e, was reached, hence for the strongest absorption of femtosecond laser pulses. The X-ray emission in Figure 2a shows the slope $\gamma = 2$ scaling for this maximized absorbed intensity. The absorbed energy density, W_{abs}, (per volume) is the relevant quantity that should be considered, since the free electron density, n_e, is approaching the critical density, n_{cr}, (plasma reflection range) during the light absorption and $W_{abs} \propto \frac{n_e}{n_{cr}} F_p$, where F_p is the pulse fluence [43]. The absorbed energy density was defined by the mechanism of electron generation $n_e \propto I_p^m \propto F_p^m$ and determined by the corresponding exponent m ($m = 1$ for the linear absorption). The strongest absorption took place when permittivity $\epsilon \Rightarrow 0$ when material was transforming from dielectric to metal-like [50]. Under these conditions, direct light absorption $m = 1$ was dominant regardless of the initial nonlinearity required to excite free carriers and to decrease the real part of the permittivity. The experimentally observed slope $\gamma = 2$ shown in Figure 2a of the X-ray intensity as a function of the excitation laser intensity energy E_p (or I_p, F_p) was expected. This signifies a decreasing volume where light was absorbed at the increasing electronic excitation. This was confirmed by femtosecond laser ablation and etching where strong localization of modification took place at the very centre of the laser irradiated spot [51].

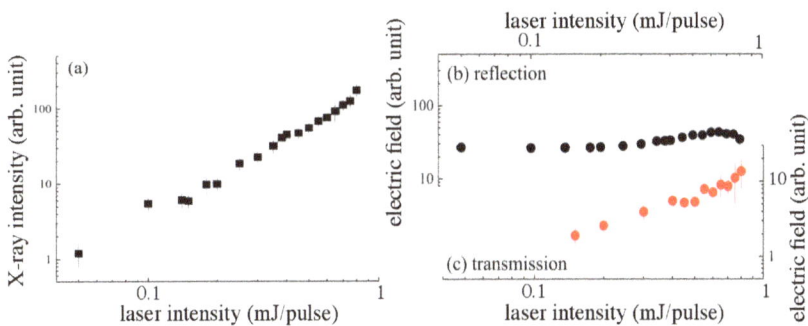

Figure 2. Laser intensity dependencies of (a) X-ray and THz wave intensities (b) in the reflection and (c) in the transmission. The laser polarization was horizontal to the solution flow, i.e., p-pol. During the measurements, the solution flow position along the Z-axis was finely optimized for the highest X-ray intensity at each laser intensity.

THz wave intensities measured in the reflection and the transmission are shown in Figure 2b,c, respectively. For the 20 μm-thick solution flow, it showed different behaviours: THz wave emission in the reflection was almost independent of the pulse intensity, while that in transmission was linearly increasing $E_{THz} \propto E_p$ for the p-polarized laser excitation. At the used $NA = 0.25$, the geometric focus diameter was $d = 1.22\lambda/NA \sim 3.9$ μm, which for the smallest excitation laser intensity $E_p = 0.1$ mJ defined the intensity $I_p = E_p/(\pi(d/2)^2 t_p) = 24$ PW/cm^2. The actual pulse intensity was considerably lower than this calculation due to air breakdown and intensity clamping known in the filamentation of femtosecond pulses [43]. Such high irradiance conditions are usually not explored in the studies of THz wave emission, and low irradiance scaling is usually $P_{THz} \propto P_I^2$ [52].

There have been controversial and various discussions on THz wave emission mechanisms from laser-induced plasmas. As the first step for such discussions on water, however, it is meaningful to learn discussions on similar experimental conditions with metal targets [52] to those in this paper with water; under a grazing angle laser excitation, with flat solution flow and the THz wave radiation from the flow surface. THz wave emission from solid flat surfaces under a grazing angle excitation, α, (angle between the surface and the laser beam) was usually found to be much longer at ~ 1 ps (1 THz) than the excitation ultra-short laser pulse of tens of femtoseconds; the angle of incidence was ($\alpha \sim \pi/2$) [53]. The excitation laser pulse was travelling on the surfaces at the velocity $v = c/\cos\alpha$, and the heated electrons emitted the THz field $E_{THz} \propto \partial T_e/\partial t \propto E_\perp/\sin\alpha$, where $E_\perp$ is the light field component perpendicular to the solid surface (p-component of the laser E-field), T_e is the electron temperature and c and t are the speed of light and time, respectively [52]. This Cherenkov-type synchronism [52] explains THz wave emission in specular reflection, its polarization (defined by $E_\perp$)

and the angular dependence of the THz wave emission pattern on the light incidence angle for the *s* and *p*−polarisations. The electron scattering mechanisms and their temperature dependence define the strongly nonlinear scaling of the emitted THz power $P_{THz} \propto P_l^{4/(2-n)}$, where P_l is the laser pulse power and *n* ($n < 2$) is the exponent defining the temperature dependence of the effective electron collision frequency $\nu = \nu_0 + \beta T^n$ [52]. Arguably, in the case of water, the plasma breakdown on its surface acts like a metal mirror for the THz wave generated in the air breakdown region, as well as from a plasma skin depth travelling on the flow surface at the slanted irradiation, as discussed above (see the inset in Figure 1a). The transmission of the THz wave was weaker, but was linearly increasing with the excitation laser intensity, as shown in Figure 2c. This is an indication of either a larger excited volume or a higher temperature of electrons (considering that there was no strong difference in electron scattering behaviour). The larger excited volume emitting the THz wave was one probable cause, since the dependence of the THz wave emission on the axial position of optical femtosecond laser excitation on the solution flow showed a wider axial width, as discussed next.

3.2. Z-Scan for THz Wave and X-ray Emission

Figure 3 shows the X-ray and THz wave intensity profiles at different solution positions along the Z-axis when the excitation laser intensities were 0.2 and 0.7 mJ/pulse. The X-ray profile (Figure 3a) showed an asymmetric feature with a longer tail to the upstream side. This reflects that the X-ray emission mechanism was mainly related to laser-induced plasma formation [25]. When the solution was set at the downstream side, the incident laser light was partly reflected by the self-induced plasma, which resulted in the instantaneous degradation of X-ray intensity observed in the downstream side. When the excitation laser intensity increased to 0.7 mJ/pulse (Figure 3c), the profile width became broader from 44 μm to 50 μm with a broader tail, and this may indicate that the X-ray source size became enlarged. X-ray emission spectra in the hard X-ray region from water under these experimental conditions [38,39] showed a broadband emission due to bremsstrahlung with no characteristic lines since such bands of oxygen and hydrogen are far in the longer wavelength region [54] and the estimated electron temperature, T_e, was up to 2 keV at the highest [38,39].

The peak position of THz wave emission in the reflection direction (Figure 3b) was almost the same as that of X-ray emission though its width of THz wave emission was apparently broader at 225 μm compared with X-ray emission. This implies that there is a mechanism to enhance the THz emission from the water plasma with low electron density and temperatures when the focal point is far from the water film. One possible mechanism is the four-wave mixing/optical rectification process [17] in which the second harmonic (400 nm) component of the white light continuum was generated in air at the focal point by self-phase modulation and the residual excitation laser pulse was mixed, which resulted in the enhancement of THz wave emission from the water plasma at the far sides from the peak position. On the other hand, the profile width when the laser intensity was 0.7 mJ/pulse (Figure 3d) became as narrow as the width of the X-ray emission, which indicates that the mechanism of laser-induced plasma formation dominated more in THz wave emission when the laser intensity increased. THz wave emission from only air was negligibly small; however, when the solution position was set far from the laser focus $z > 200$ μm at the downstream side, the emission became relatively dominant, as shown in Figure 3b.

In the transmission, THz wave emission changed its nature from that in the reflection. When the laser intensity was 0.2 mJ/pulse, intensity degradation at the peak position for THz wave emission in the reflection was clearly observed (Figure 3b). In addition, local peaks at the upstream side ($z = -135$ μm) and the downstream side ($z = 24$ μm) were observed as indicated by red arrows. This feature may be assigned to the laser-induced plasma especially, at Z-positions close to the THz wave emission peak in reflection; the laser-induced plasma reflected THz wave radiation. Under this hypothesis, at the positions indicated by the red arrows, the plasma density reached its critical density for THz wave, defined as $\omega_p = \sqrt{n_e e^2 / m_e \varepsilon_0}$, where ω_p, n_e, e, m_e and ε_0 are the plasma cyclic frequency, electron density, its mass and permittivity. The critical electron density $n_e = n_{cr}$ was

1.23×10^{16} cm^{-3} for 1 THz. The Debye length, λ_D, could be also estimated to be about 2.3 µm with T_e at 2 keV [38,39]. Similarly, when the laser pulse energy increased to 0.7 mJ/pulse (Figure 3d), such features were clearly observed, and the position range along the Z-axis with the plasma density higher than critical $n_{cr} = 1.23 \times 10^{16}$ cm^{-3} was changed from $\Delta z_1 = 159$ µm (0.2 mJ/pulse) to $\Delta z_2 = 229$ µm (0.7 mJ/pulse) along the Z-axis. As in the discussion of Figure 3b described above, the broad tails at the upstream and the downstream sides can be assigned to optical coherent processes such as four-wave mixing. Its relative intensity at the tails to the intensity close to zero-position was increasing if compared to the case with 0.2 mJ/pulse (Figure 3b), which was considered to be the result of the dominant plasma effect at the zero-position when the laser intensity was higher. Furthermore, similarly to the case of the reflection shown in Figure 3b, THz wave emission from air became dominant when the solution position was far from the laser focus and was observed as the constant THz wave intensity at the solution position $z > 250$ µm at the downstream side. For all the excitation laser intensities, the high reflectivity correlated with low transmittance of the water film target.

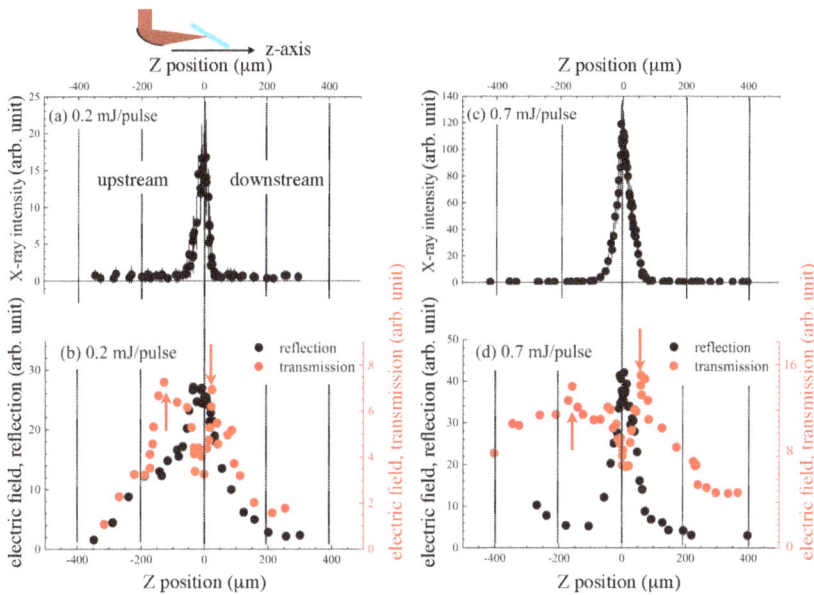

Figure 3. X-ray (**a**) and THz wave (**b**) intensities as a function of the flow position along the Z-axis when the laser intensity was 0.2 mJ/pulse and 0.7 mJ/pulse (**c,d**); Red arrows in (**b,d**) for the transmission indicate the Z-positions at (**b**) -135 µm and 24 µm ($\Delta z_1 = 159$ µm) and (**d**) -167 µm and 62 µm ($\Delta z_2 = 229$ µm), respectively.

3.3. THz Wave Intensity Enhancements under Double-Pulse Excitation

One experiment for THz wave emission enhancements was performed under a double-pulse excitation. Figure 4 shows X-ray emission (a) and THz wave emission (b,c) with the pre-pulse (0.1 mJ/pulse, vertically-polarized, s-pol.) with the delay time of 4.6 ns in advance of the main pulse (0.4 mJ/pulse, horizontally-polarized, p-pol.) irradiation. X-ray emission apparently showed an intensity enhancement under the double-pulse excitation, as expected [37,38]. One additional peak at the downstream side was also clearly discernible, as reported recently [43]. With a time delay of 4.6 ns between the pulses, the initial processes of water film ablation and transient surface roughening under action of capillary forces and micro-droplet formation (mist) at the close position to the initial location of the solution surface, all induced by the pre-pulse irradiation, caused a more effective coupling of the main pulse with such a modified solution surface. The enhancement was caused by multiple scattering, local refocusing of light by droplets and the perturbed surface, which resulted in the X-ray intensity enhancement, which can reach an order of magnitude and is useful for practical applications. THz wave emission in the reflection (Figure 4b) was also enhanced about five times, and the profile width under the double-pulse excitation became narrower at 52 µm as compared with the 106 µm width under the single-pulse excitation. The profile of THz wave emission along the Z-axis under the double-pulse excitation showed only a single peak at the centre as in the case of the single-pulse excitation, which was different from the profile of X-ray emission with the additional peak at the downstream side. This is consistent with requirement of thermal gradients lasting ~1 ps for ~1 THz emission, which are less likely on a fragmented water film, while X-ray emission is maintained by hot plasma and geometrical factors are less important. A detailed investigation is needed using time-resolved shadowgraphy to reveal the geometrical evolution of the disintegrating water surface. In the case of THz wave emission in the transmission, the profile showed an apparent change (Figure 4c), namely the intensity was enhanced more than ten times, and the profile along the Z-axis changed to a single peak from the profiles with a dip at its centre, as shown in Figure 3b,d. Under the pre-pulse irradiation condition, transient surface roughness, droplet (mist) formation and hole formation on the solution flow were expected at a delay time of 4.6 ns [37]. These initial processes of laser ablation induced by the pre-pulse irradiation may cause the enhancements of THz wave emission especially in the transmission direction.

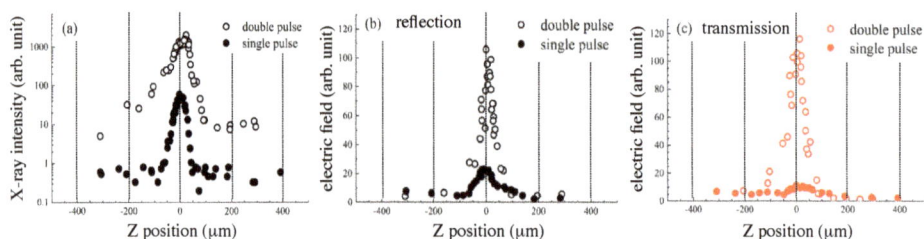

Figure 4. Z-position-dependent intensities of the X-ray in log-scale (**a**), the THz wave in the reflection (**b**) and the transmission (**c**) under the double-pulse excitation condition. Filled circles are under the single-pulse excitation condition, and open circles are under the double-pulse excitation condition, where the laser intensities for the main excitation pulse (horizontally-polarized, p-pol.) and the pre-pulse (vertically-polarized, s-pol., 4.6 ns in advance of the main pulse) were 0.4 mJ/pulse and 0.1 mJ/pulse, respectively.

4. Conclusions and Outlook

This study reports the demonstration of the dual X-ray and THz wave simultaneous emission from water flow irradiated by focused femtosecond laser pulses in air. Different characteristic features of THz wave emission, which are associated with X-ray emission, in the reflection and the transmission under the single-pulse excitation were clearly revealed. Enhancements of THz wave emission under the double-pulse excitation up to 5–10 times were also shown and indicated that further enhancements of THz wave emission are expected by optimizing the laser excitation conditions. Another option as laser parameters to enhance THz wave emission is laser chirp [24] or double-colour excitation with the fundamental and the second harmonic, expecting efficient augmentation of energy delivery to the target via an optical process such as four-wave-mixing [17]. Various solution samples can be utilised as targets, since X-ray emission from water is also enhanced by the addition of electrolyte [39] and gold nano-particles [40]. Other materials such as bismuth [55] and copper [56] can be also used as targets at the tested high-irradiance conditions. Interaction of the X-ray or THz wave with matter originates with electrons at keV or meV, in other words with electrons bound in inner-shells or with structural absorption resonances, respectively. Based on the basics, X-ray and THz wave science and technology for spectroscopy and imaging have made their progress independently [57,58]. Under experiments in water or at atmospheric pressure, ultrasound emission is also expected [59], and super-resolution photoacoustic imaging is also further developed [60]. With synchronized X-ray and THz wave emission as introduced in this paper, not only for the basic mechanism study on THz wave emission from aqueous solutions based on laser-plasma dynamics, but also combined synchronous usages of X-ray and THz wave or ultrasound are expected to contribute well to studies on nanomaterials from the nano-scale viewpoints to macro-scales.

Author Contributions: H.-h.H., T.N. and K.H. performed the experiments. W.-h.H. was involved in the system setup. H.-h.H., T.N., S.J. and K.H. analysed the data and wrote the paper.

Acknowledgments: K.H. acknowledges the Japan Science and Technology Agency (JST) PRESTO (Precursory Research for Embryonic Science and Technology) Program (SAKIGAKE, Innovative use of light and materials/life) for its partial support of this research. S.J. is grateful for the support via the Australian Research Council DP170100131 grant.

Funding: This research received no external funding.

Conflicts of Interest: The authors declare no conflict of interest. The founding sponsors had no role in the design of the study; in the collection, analyses or interpretation of data; in the writing of the manuscript; nor in the decision to publish the results.

References

1. Bagratashvili, V.N.; Letokhov, V.S.; Makarov, A.A.; Ryabov, E.A. (Eds.) *Multiple Photon Infrared Laser Photophysics and Photochemistry*; Harwood Academic Publishers: Reading, UK, 1985.
2. Nakajima, K.; Deguchi, M. (Eds.) *Science of Superstrong Field Interactions*; American Institute of Physics: College Park, MD, USA, 2002.
3. Yamanouchi, K.; Midorikawa, K. (Eds.) *Multiphoton Processes and Attosecond Physics*; Springer: New York, NY, USA, 2012.
4. Corde, S.; Phuoc, K.T.; Lambert, G.; Fitour, R.; Malka, V.; Rousse, A.; Beck, A.; Lefebvre, E. Femtosecond X rays from laser-plasma accelerators. *Rev. Mod. Phys.* **2013**, *85*, 1–48. [CrossRef]
5. Dey, I.; Jana, K.; Fedorov, V.Y.; Koulouklidis, A.D.; Mondal, A.; Shaikh, M.; Sarkar, D.; Lad, A.D.; Tzortzakis, S.; Couairon, A.; et al. Highly efficient broadband terahertz generation from ultrashort laser filamentation in liquids. *Nat. Commun.* **2017**, *8*, 1184. [CrossRef] [PubMed]
6. Fork, R.L.; Shank, C.V.; Hirlimann, C.; Yen, R.; Tomlinson, W.J. Femtosecond white-light continuum pulses. *Opt. Lett.* **1983**, *8*, 1–3. [CrossRef] [PubMed]
7. Rethfeld, B.; Ivanov, D.S.; Garcia, M.E.; Anisimov, S.I. Modelling ultrafast laser ablation. *J. Phys. D Appl. Phys.* **2017**, *50*, 193001. [CrossRef]
8. Helliwell, J.R.; Rentzepis, P.M. (Eds.) *Time-Resolved Diffraction*; Oxford Science Publications: Oxford, UK, 1986.
9. Hoffmann, M.C.; Fülöp, J.A. Intense ultrashort terahertz pulses: Generation and applications. *J. Phys. D Appl. Phys.* **2011**, *44*, 083001. [CrossRef]
10. Tonouchi, M. Cutting-edge terahertz technology. *Nat. Photonics* **2007**, *1*, 97–105. [CrossRef]
11. Kampfrath, T.; Tanaka, K.; Nelson, K.A. Resonant and nonresonant control over matter and light by intense terahertz transients. *Nat. Photonics* **2013**, *7*, 680–690. [CrossRef]
12. Wenz, J.; Schleede, S.; Khrennikov, K.; Bech, M.; Thibault, P.; Heigoldt, M.; Pfeiffer, F.; Karsch, S. Quantitative X-ray phase-contrast microtomography from a compact laser-driven betatron source. *Nat. Commun.* **2015**, *6*, 7568. [CrossRef] [PubMed]
13. Baierl, S.; Hohenleutner, M.; Kampfrath, T.; Zvezdin, A.K.; Kimel, A.V.; Huber, R.; Mikhaylovskiy, R.V. Nonlinear spin control by terahertz-driven anisotropy fields. *Nat. Photonics* **2016**, *10*, 715–718. [CrossRef]
14. Phillips, K.C.; Gandhi, H.H.; Mazur, E.; Sundaram, S.K. Ultrafast laser processing of materials: A review. *Adv. Opt. Photonics* **2015**, *7*, 684–712. [CrossRef]
15. Hamster, H.; Sullivan, A.; Gordon, S.; White, W.; Falcone, R.W. Subpicosecond, electromagnetic pulses from intense laser-plasma interaction. *Phys. Rev. Lett.* **1993**, *71*, 2725–2728. [CrossRef] [PubMed]

16. Löffler, T.; Jacob, F.; Roskos, H.G. Generation of terahertz pulses by photoionization of electrically biased air. *Appl. Phys. Lett.* **2000**, *77*, 453–455. [CrossRef]

17. Cook, D.J.; Hochstrasser, R.M. Intense terahertz pulses by four-wave rectification in air. *Opt. Lett.* **2000**, *25*, 1210–1212. [CrossRef] [PubMed]

18. D'Amico, C.; Houard, A.; Franco, M.; Prade, B.; Mysyrowicz, A.; Couairon, A.; Tikhonchuk, V.T. Conical forward THz emission from femtosecond-laser-beam filamentation in air. *Phys. Rev. Lett.* **2007**, *98*, 235002. [CrossRef] [PubMed]

19. Nagashima, T.; Hirayama, H.; Shibuya, K.; Hangyo, M.; Hashida, M.; Tokita, S.; Sakabe, S. Terahertz pulse radiation from argon clusters. *Opt. Express* **2009**, *17*, 8907. [CrossRef] [PubMed]

20. Balakin, A.V.; Dzhidzhoev, M.S.; Gordienko, V.M.; Esaulkov, M.; Zhvaniya, I.A.; Ivanov, K.A.; Kotelnikov, I.; Kuzechkin, N.A.; Ozheredov, I.A.; Panchenko, V.Y.; et al. Interaction of high-intensity femtosecond radiation with gas cluster beam: Effect of pulse duration on joint terahertz and X-ray emission. *IEEE Trans. Terahertz Sci. Technol.* **2016**, *7*, 70–79. [CrossRef]

21. Mori, K.; Hashida, M.; Nagashima, T.; Li, D.; Teramoto, K.; Nakamiya, Y.; Inoue, S.; Sakabe, S. Directional linearly polarized terahertz emission from argon clusters irradiated by noncollinear double-pulse beams. *Appl. Phys. Lett.* **2017**, *111*, 241107. [CrossRef]

22. Sagisaka, A.; Daido, H.; Nashima, S.; Orimo, S.; Ogura, K.; Mori, M.; Yogo, A.; Ma, J.; Daito, I.; Pirozhkov, A.S.; et al. Simultaneous generation of a proton beam and terahertz radiation in high-intensity laser and thin-foil interaction. *Appl. Phys. B* **2008**, *90*, 373–377. [CrossRef]

23. Tokita, S.; Sakabe, S.; Nagashima, T.; Hashida, M.; Inoue, S. Strong sub-terahertz surface waves generated on a metal wire by high-intensity laser pulses. *Sci. Rep.* **2015**, *5*, 8268. [CrossRef] [PubMed]

24. Qi, J.; Yiwen, E.; Williams, K.; Dai, J.; Zhang, X.-C. Observation of broadband terahertz wave generation from liquid water. *Appl. Phys. Lett.* **2017**, *111*, 071103.

25. Attwood, D. *Soft X-rays and Extreme Ultraviolet Radiation*; Cambridge University Press: Cambridge, UK, 1999.

26. Turcu, I.C.E.; Dance, J.B. *X-rays from Laser Plasmas*; WILEY: Hoboken, NJ, USA, 1998.

27. Yoshida, M.; Fujimoto, Y.; Hironaka, Y.; Nakamura, K.G.; Kondo, K.; Ohtani, M.; Tsunemi, H. Generation of picosecond hard x rays by tera watt laser focusing on a copper target. *Appl. Phys. Lett.* **1998**, *73*, 2393–2395. [CrossRef]

28. Hatanaka, K.; Yomogihata, K.; Ono, H.; Nagafuchi, K.; Fukumura, H.; Fukushima, M.; Hashimoto, T.; Juodkazis, S.; Misawa, H. Hard X-ray generation using femtosecond irradiation of PbO glass. *J. Non-Cryst. Solids* **2008**, *354*, 5485–5490. [CrossRef]

29. Hansson, B.; Rymell, L.; Berglund, M.; Hertz, H. A liquid-xenon-jet laser-plasma x-ray and EUV source. *Microelectron. Eng.* **2000**, *53*, 667–670. [CrossRef]

30. Vogt, U.; Stiel, H.; Will, I.; Nickles, P.V.; Sandner, W.; Wieland, M.; Wilhein, T. Influence of laser intensity and pulse duration on the extreme ultraviolet yield from a water jet target laser plasma. *Appl. Phys. Lett.* **2001**, *79*, 2336–2338. [CrossRef]

31. Düsterer, S.; Schwoerer, H.; Ziegler, W.; Ziener, C.; Sauerbrey, R. Optimization of EUV radiation yield from laser-produced plasma. *Appl. Phys. B* **2001**, *73*, 693–698. [CrossRef]

32. Hansson, B.A.M.; Berglund, M.; Hemberg, O.; Hertz, H.M. Stabilization of liquified-inert-gas jets for laser–plasma generation. *J. Appl. Phys.* **2004**, *95*, 4432–4437. [CrossRef]

33. Rajyaguru, C.; Higashiguchi, T.; Koga, M.; Sasaki, W.; Kubodera, S. Systematic optimization of the extreme ultraviolet yield from a quasi-mass-limited water-jet target. *Appl. Phys. B* **2004**, *79*, 669–672. [CrossRef]

34. Hansson, B.A.M.; Hemberg, O.; Hertz, H.M.; Choi, M.B.H.J.; Jacobsson, B.; Janin, E.; Mosesson, S.; Rymell, L.; Thoresen, J.; Wilner, M. Characterization of a liquid-xenon-jet laser-plasma extreme-ultraviolet source. *Rev. Sci. Instrum.* **2004**, *75*, 2122–2129. [CrossRef]

35. Hsu, W.H.; Masim, F.C.P.; Porta, M.; Nguyen, M.T.; Yonezawa, T.; Balčytis, A.; Wang, X.; Rosa, L.; Juodkazis, S.; Hatanaka, K. Femtosecond laser-induced hard X-ray generation in air from a solution flow of Au nano-sphere suspension using an automatic positioning system. *Opt. Express* **2016**, *24*, 19994–20001. [CrossRef] [PubMed]

36. Hatanaka, K.; Miura, T.; Fukumura, H. White X-ray pulse emission of alkali halide aqueous solutions irradiated by focused femtosecond laser pulses: A spectroscopic study on electron temperatures as functions of laser intensity, solute concentration, and solute atomic number. *Chem. Phys.* **2004**, *299*, 265–270. [CrossRef]

37. Hatanaka, K.; Ono, H.; Fukumura, H. X-ray pulse emission from cesium chloride aqueous solutions when irradiated by double-pulsed femtosecond laser pulses. *Appl. Phys. Lett.* **2008**, *93*, 064103. [CrossRef]
38. Hatanaka, K.; Fukumura, H. X-ray emission from CsCl aqueous solutions when irradiated by intense femtosecond laser pulses and its application to time-resolved XAFS measurement of I in aqueous solution. *X-ray Spectrom.* **2012**, *41*, 195–200. [CrossRef]
39. Hatanaka, K.; Miura, T.; Fukumura, H. Ultrafast X-ray pulse generation by focusing femtosecond infrared laser pulses onto aqueous solutions of alkali metal chloride. *Appl. Phys. Lett.* **2002**, *80*, 3925–3927. [CrossRef]
40. Masim, F.C.P.; Porta, M.; Hsu, W.H.; Nguyen, M.T.; Yonezawa, T.; Balčytis, A.; Juodkazis, S.; Hatanaka, K. Au nanoplasma as efficient hard X-ray emission source. *ACS Photonics* **2016**, *3*, 2184–2190. [CrossRef]
41. Zhang, X.C.; Shkurinov, A.; Zhang, Y. Extreme terahertz science. *Nat. Photonics* **2017**, *11*, 16–18. [CrossRef]
42. Hatanaka, K.; Ida, T.; Ono, H.; Matsushima, S.; Fukumura, H.; Juodkazis, S.; Misawa, H. Chirp effect in hard X-ray generation from liquid target when irradiated by femtosecond pulses. *Opt. Express* **2008**, *16*, 12650–12657. [CrossRef] [PubMed]
43. Hsu, W.H.; Masim, F.C.P.; Balčytis, A.; Juodkazis, S.; Hatanaka, K. Dynamic position shifts of X-ray emission from a water film induced by a pair of time-delayed femtosecond laser pulses. *Opt. Express* **2017**, *25*, 24109–24118. [CrossRef] [PubMed]
44. Hatanaka, K.; Tsuboi, Y.; Fukumura, H.; Masuhara, H. Nanosecond and Femtosecond Laser Photochemistry and Ablation Dynamics of Neat Liquid Benzenes. *J. Phys. Chem. B* **2002**, *106*, 3049–3060. [CrossRef]
45. Wu, Q.; Zhang, X. Free-space electro-optic sampling of terahertz beams. *Appl. Phys. Lett.* **1995**, *67*, 3523–3525. [CrossRef]
46. Lee, Y.S. *Principles of Terahertz Science and Technology*; Springer: New York, NY, USA, 2009.
47. Exter, M.V.; Fattinger, C.; Grischkowsky, D. Terahertz time-domain spectroscopy of water vapour. *Opt. Lett.* **1989**, *14*, 1128–1130. [CrossRef] [PubMed]
48. Novelli, F.; Chon, J.W.M.; Davis, J.A. Terahertz thermometry of gold nanospheres in water. *Opt. Lett.* **2016**, *41*, 5801–5804. [CrossRef] [PubMed]
49. Thrane, L.; Jacobsen, R.H.; Uhd Jepsen, P.; Keiding, S.R. THz reflection spectroscopy of liquid water. *Chem. Phys. Lett.* **1995**, *240*, 330–333. [CrossRef]
50. Gamaly, E.G.; Rode, A.V. Ultrafast re-structuring of the electronic landscape of transparent dielectrics: New material states (Die-Met). *Appl. Phys. A* **2018**, *124*, 278. [CrossRef]
51. Cao, X.W.; Chen, Q.D.; Fan, H.; Juodkazis, L.Z.S.; Sun, H.B. Liquid-Assisted Femtosecond Laser Precision-Machining of Silica. *Nanomaterials* **2018**, *8*, 287. [CrossRef] [PubMed]
52. Oladyshkin, I.V. Diagnostics of the electron scattering in metals in terms of a terahertz response to femtosecond laser pulse. *JETP Lett.* **2016**, *103*, 435–439. [CrossRef]
53. Oladyshkin, I.V.; Fadeev, D.A.; Mironov, V.A. Thermal mechanism of laser induced THz generation from a metal surface. *J. Opt.* **2015**, *17*, 075502. [CrossRef]
54. Kirz, J.; Attwood, D.; Henke, B.L.; Howells, M.R.; Kennedy, K.D.; Kim, K.J.; Kortright, J.B.; Perera, R.C.; Pianetta, P.; Riordan, J.C.; et al. *X-ray Data Booklet*; Lawrence Berkeley National Laboratory, University of California: Berkeley, CA, USA, 1986.
55. Ilyakov, I.E.; Shishkin, B.V.; Fadeev, D.A.; Oladyshkin, I.V.; Chernov, V.V.; Okhapkin, A.I.; Yunin, P.A.; Mironov, V.A.; Akhmedzhanov, R.A. Terahertz radiation from bismuth surface induced by femtosecond laser pulses. *Opt. Lett.* **2016**, *41*, 4289–4292. [CrossRef] [PubMed]
56. Lu, X.; Ishida, Y.; Mishina, T.; Nguyen, M.T.; Yonezawa, T. Enhanced Terahertz Emission from CuxO/Metal Thin Film Deposited on Columnar-Structured Porous Silicon. *Bull. Chem. Soc. Jpn.* **2015**, *88*, 1385–1387. [CrossRef]
57. Sakdinawat, A.; Attwood, D. Nanoscale X-ray imaging. *Nat. Photonics* **2010**, *4*, 840–848. [CrossRef]
58. Mittleman, D.M. Twenty years of terahertz imaging. *Opt. Express* **2018**, *26*, 9417–9431. [CrossRef] [PubMed]

59. Masim, F.C.P.; Hsu, W.H.; Liu, H.L.; Yonezawa, T.; Balčytis, A.; Juodkazis, S.; Hatanaka, K. Photoacoustic signal enhancements from gold nano-colloidal suspensions excited by a pair of time-delayed femtosecond pulses. *Opt. Express* **2017**, *25*, 19497–19507. [CrossRef] [PubMed]

60. Chaigne, T.; Arnaland, B.; Vilov, S.; Bossy, E.; Katz, O. Super-resolution photoacoustic imaging via flow-induced absorption fluctuations. *Optica* **2017**, *4*, 1397–1404. [CrossRef]

MDPI

St. Alban-Anlage 66

4052 Basel

Switzerland

Tel. +41 61 683 77 34

Fax +41 61 302 89 18

www.mdpi.com

Nanomaterials Editorial Office

E-mail: nanomaterials@mdpi.com

www.mdpi.com/journal/nanomaterials